Data Mining and Applications in Genomics

Lecture Notes in Electrical Engineering

Volume 25

For other titles published in this series, go to
www.springer.com/series/7818

Sio-Iong Ao

Data Mining and Applications in Genomics

Sio-Iong Ao
International Association of Engineers
Oxford University
UK

ISBN 978-90-481-8040-0 e-ISBN 978-1-4020-8975-6

Printed on acid-free paper

springer.com

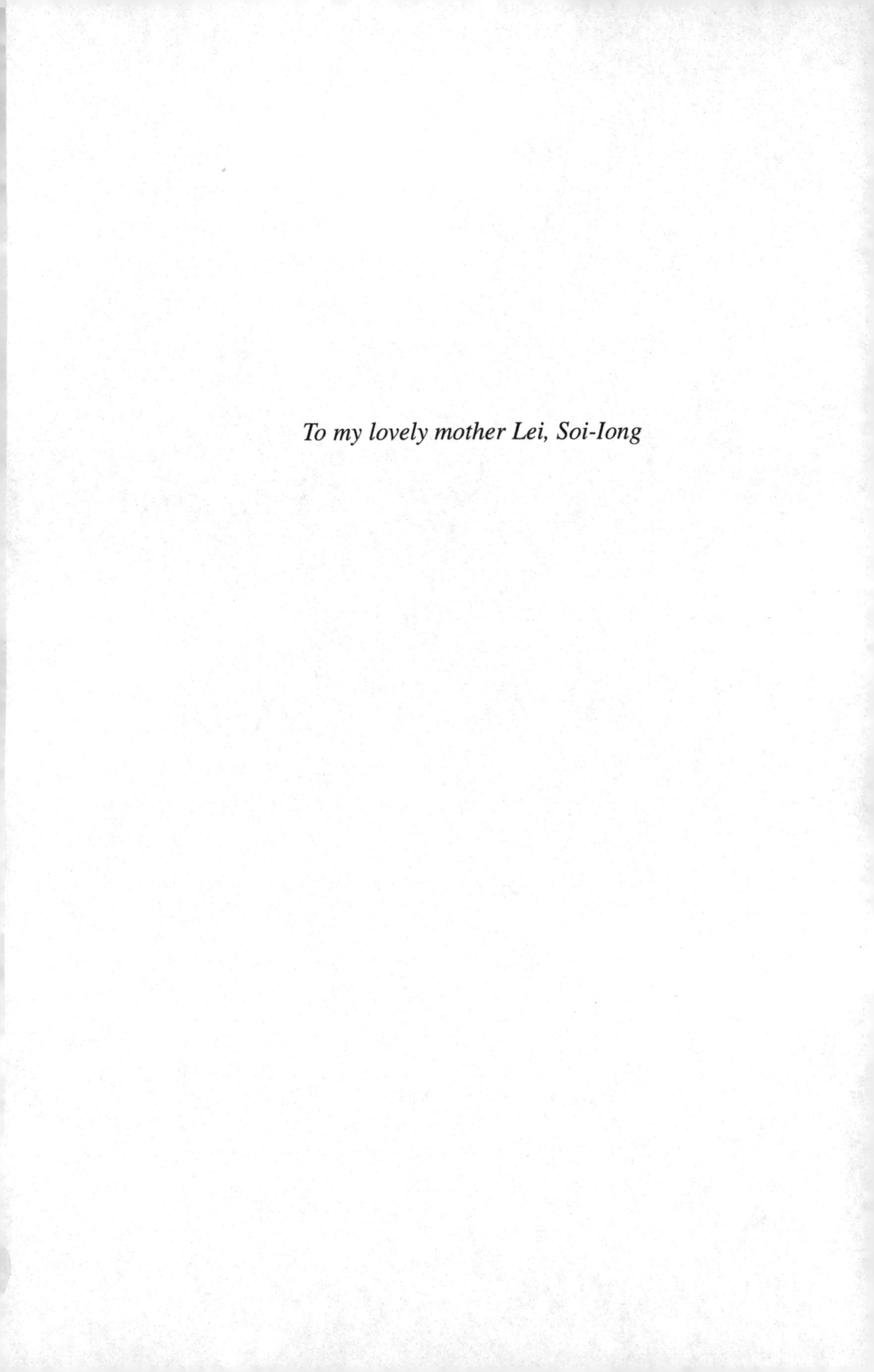

To my lovely mother Lei, Soi-Iong

Preface

With the results of many different genome-sequencing projects, hundreds of genomes from all branches of species have become available. Currently, one important task is to search for ways that can explain the organization and function of each genome. Data mining algorithms become very useful to extract the patterns from the data and to present it in such a way that can better our understanding of the structure, relation, and function of the subjects. The purpose of this book is to illustrate the data mining algorithms and their applications in genomics, with frontier case studies based on the recent and current works of the author and colleagues at the University of Hong Kong and the Oxford University Computing Laboratory, University of Oxford.

It is estimated that there exist about 10 million single-nucleotide polymorphisms (SNPs) in the human genome. The complete screening of all the SNPs in a genomic region becomes an expensive undertaking. In Chapter 4, it is illustrated how the problem of selecting a subset of informative SNPs (tag SNPs) can be formulated as a hierarchical clustering problem with the development of a suitable similarity function for measuring the distances between the clusters. The proposed algorithm takes account of both functional and linkage disequilibrium information with the asymmetry thresholds for different SNPs, and does not have the difficulties of the block-detecting methods, which can result in different block boundaries. Experimental results supported that the algorithm is cost-effective for tag-SNP selection. More compact clusters can be produced with the algorithm to improve the efficiency of association studies.

There are several different advantages of the linkage disequilibrium maps (LD maps) for genomic analysis. In Chapter 5, the construction of the LD mapping is formulated as a non-parametric constrained unidimensional scaling problem, which is based on the LD information among the SNPs. This is different from the previous LD map, which is derived from the given Malecot model. Two procedures, one with the formulation as the least squares problem with nonnegativity and the other with the iterative algorithms, have been considered to solve this problem. The proposed maps can accommodate recombination events that have accumulated. Application of the proposed LD maps for human genome is presented. The linkage disequilibrium patterns in the LD maps can provide the genomic information like the hot and cold recombination regions, and can facilitate the study of recent selective sweeps across the human genome.

Microarray has been the most widely used tool for assessing differences in mRNA abundance in the biological samples. Previous studies have successfully employed principal components analysis-neural network as a classifier of gene types, with continuous inputs and discrete outputs. In Chapter 6, it is shown how to develop a hybrid intelligent system for testing the predictability of gene expression time series with PCA and NN components on a continuous numerical inputs and outputs basis. Comparisons of results support that our approach is a more realistic model for the gene network from a continuous prospective.

In this book, data mining algorithms have been illustrated for solving some frontier problems in genomic analysis. The book is organized as follows. In Chapter 1, it is the brief introduction to the data mining algorithms, the advances in the technology and the outline of the recent works for the genomic analysis. In Chapter 2, we describe about the data mining algorithms generally. In Chapter 3, we describe about the recent advances in genomic experiment techniques. In Chapter 4, we present the first case study of CLUSTAG & WCLUSTAG, which are tailor-made hierarchical clustering and graph algorithms for tag-SNP selection. In Chapter 5, the second case study of the non-parametric method of constrained unidimensional scaling for constructions of linkage disequilibrium maps is presented. In Chapter 6, we present the last case study of building of hybrid PCA-NN algorithms for continuous microarray time series. Finally, we give the conclusions and some future works based on the case studies in Chapter 7.

Topics covered in the book include Genomic Techniques, Single Nucleotide Polymorphisms, Disease Studies, HapMap Project, Haplotypes, Tag-SNP Selection, Linkage Disequilibrium Map, Gene Regulatory Networks, Dimension Reduction, Feature Selection, Feature Extraction, Principal Component Analysis, Independent Component Analysis, Machine Learning Algorithms, Hybrid Intelligent Techniques, Clustering Algorithms, Graph Algorithms, Numerical Optimization Algorithms, Data Mining Software Comparison, Medical Case Studies, Bioinformatics Projects, and Medical Applications etc. The book can serve as a reference work for researchers and graduate students working on data mining algorithms and applications in genomics.

The author is grateful for the advice and support of Dr. Vasile Palade throughout the author's research in Oxford University Computing Laboratory, University of Oxford, UK.

June 2008 Sio-Iong Ao
University of Oxford, UK

Contents

Chapter 1
Introduction

This book is organized as follows. In this Chapter, it is the brief introduction to the data mining algorithms, the advances in the technology and the outline of the recent works for the genomic analysis. In the last section, we describe briefly about the three case studies of developing tailor-made data mining algorithms for genomic analysis. The contributions of these algorithms to the genomic analysis are also described briefly in that section and in more details in their respective case study chapters. In Chapter 2, we describe about the data mining algorithms generally. In Chapter 3, we describe about the recent advances in genomic experiment techniques. In Chapter 4, we present the first case study of CLUSTAG & WCLUSTAG, which are tailor-made hierarchical clustering and graph algorithms for tag-SNP selection. In Chapter 5, the second case study of the non-parametric method of constrained unidimensional scaling for constructions of linkage disequilibrium maps is presented. In Chapter 6, we present the last case study of building of hybrid PCA-NN algorithms for continuous microarray time series. Finally, we give the conclusions and some future works based on the case studies in Chapter 7.

1.1 Data Mining Algorithms

1.1.1 Basic Definitions

Data mining algorithm has played an important role in the overall knowledge-discovery process. It usually involves the following steps (Bergeron, 2003):

1. To select enough sample data form the sources.
2. To preprocess and clean the data, for removing errors and redundancies.
3. To transform or reduce the data to a space more suitable for data mining.
4. To implement the data mining algorithms.
5. To evaluate the mined data.
6. To present the evaluation results in a format/graph that can be understand easily.
7. To design new data queries for testing new hypotheses and return to step 1.

The above procedure of the knowledge-discovery process is in fact an iterative process that involves feedback at each stage. These feedbacks can be made within the algorithms or, made by human experts. For example, if it is the preprocessing and cleaning of a microarray dataset that cause insufficient number of records, the researcher may need to re-formulate the selection and sampling requirement for larger number of records. In some worse cases, one may even need to increase the number of microarray experiments.

Not only helpful for large datasets, data mining algorithms can be applied for relatively small datasets too. For example, in a microarray experiment, which may only have a few subjects and a few hundred records for each subject, the data mining algorithms can assist us to find out joint hypotheses, like the combination of several records. The algorithms can also search through more subjects, more records and more genomic regions, which may otherwise become much more labor-intensive or even infeasible to do it manually. Tailor-made data mining algorithms are developed to serve these purposes in a fast and efficient way, as an alternative to manual searching.

1.1.2 Basic Data Mining Techniques

Different data mining algorithms like unsupervised learning (clustering), supervised learning (classification), regression, and machine learning techniques etc., can be employed to extract or mine meaningful patterns from the data (Bergeron, 2003). Clustering algorithms can group data into similar groups without any pre-defined classes, and we will discuss about its application for the genomic study in more details later. Classification involves the task of assigning class labels to different data records. The classification rule can be based on the minimum proximity to the center of a particular class etc. In regression methods, numerical values are assigned to the data, basing on some pre-defined statistical functions. A simple case is the linear regression of the form: $y = mx + b$. More complex functions like non-linear functions can be adapted too, which may reflect the underlying properties of the data better than the simplified linear case.

In Chapter 2, we will describe about several groups of the basic data mining algorithms. In the section of dimension reduction and transformation, we will talk about the feature selection, feature extraction methods like principal component analysis and independent component analysis etc. In the section of machine learning algorithms, topics like logistic regression models, neural network models, fuzzy systems, ensemble methods, support vector machines and hybrid intelligent techniques etc. will be covered. Then, we will also talk about the clustering algorithms like hierarchical clustering, partition clustering spectral clustering, and their considerations. In the section of graph algorithms, topics like computer representations of graphs, breadth-first search algorithms, and depth-first search algorithms will be covered. Chapter 2 will conclude with the discussion of several popular numerical optimization algorithms like steepest descent method, Newton's method, sequential unconstrained minimization, reduced gradient methods, and interior-point methods.

1.1.3 Computational Considerations

Data mining algorithms are computational algorithms that can deal with a large amount of data, and, that are capable of solving complex problems. An algorithm is a precisely defined procedure for solving a well-defined problem (Salzberg et al., 1998). In other words, an algorithm is a finite sequence of logical and mathematical instructions for the solution of a given well-defined problem (Foulds, 1991). A useful algorithm has the following characteristics: finiteness, definiteness (without ambiguity), input, output and effectiveness. An algorithm can be specified with a word statement, a list of mathematical steps, a flow chart or a computational program. A computational program refers to the embodiment of such an algorithm. During the algorithm designing for the problem solving, there are important factors that need thorough consideration. Among these, the computing time and memory space requirement are two major factors.

The speed of a computational algorithm can be measured by how many operations it needs to run. An operation is defined as a primitive machine-level instruction. Or, with some high-level abstraction, it can be defined as, for example, a single retrieval from the database etc. The definition depends on the nature of the problem for counting convenience. For example, in the protein comparison program BLAST, which requires the comparison of amino acids against each other, one operation may be defined as the comparing of one amino acid to another one. Then, one operation will require fetching two memory locations and using them as indices into a PAM matrix.

The number of operations required usually increases with the size of the input N. For example, in the sequence comparison, it will take longer to compute with longer sequence. In this sequence comparison case, we can set the input size N as the sum of the length of the sequences. For describing the computation time of an algorithm, we can say that it takes N units of time, or N^2, or maybe N^3 etc. With the concept of the operation and its counting with input size, we can have a machine-independent comparison of the computational durations of different algorithms. The notations like $O(N)$, $O(N^2)$ or $O(N^3)$ are usually used to denote the order of the number of operations an algorithm needs.

The complexity in the above paragraph refers to the maximum number of operation steps required by an algorithm, with the consideration of all possible problem instances of a given problem size. This is called the worst-case complexity. Another popular kind of complexity is the expected time complexity (average time complexity). For a given problem that is solved by two different algorithms, the complexities of these two algorithms can be different, for example the first algorithm with $O(N)$ while the second algorithm with $O(N^2)$. Then, the first algorithm is said to be more efficient than the second one.

The space requirement of an algorithm is also a function of the input size N. As an example, in the Smith-Waterman sequence comparison algorithm, a matrix of the two input sequences is built. Let N and M denote the sizes of these two sequences, the matrix is of size $N \times M$, in which each entry is consisted of a number plus a pointer. We say that the space requirement of this algorithm is of order $O(NM)$.

Sometimes, it is possible to reduce the space requirement by scarifying the running time of the algorithm. For example, with an alternative programming of the Smith-Waterman algorithm of about two times the original running time, the space requirement can be lower to O(*N*) (Waterman, 1995).

1.2 Advances in Genomic Techniques

The genomic analysis is concerned with the different properties and the variations of the genome, and it is usually in a scale much larger than the traditional genetic studies done before. There are different approaches for the genomic analysis, like the comparisons of gene order, codon usage bias, and GC-content etc. In genomic analysis, there have been advances in the technology for DNA sequencing, and, in the routine adoption of the DNA microarray technology for the analysis of gene expression profiles at the mRNA level (Lee and Lee, 2000). There have also been advances for the genotyping of single nucleotide polymorphisms in the human genome.

A collection of DNA of which an organism consists is called a genome (Pevsner, 2003). It is the genome that contains the hereditary information of that organization. This term was first used by Professor Hans Winkler of University of Hamburg in 1920 (PloS, 2005). The genome of an organism includes both the genes and the non-coding sequences. Both the genes and the other DNA elements together define the genome's identity. The sizes of the genomes can vary hugely among different species. For example, the smallest viruses have fewer than 10 genes, but in human genome, there are billions of base pair of DNA that encode tens of thousands genes.

With the results of many different genome-sequencing projects, hundreds of genomes from all branches of species have become available now. Nowadays, one important task is to search for methods that can explain each genome's organization and function. This process will need algorithms and tools from computer science, statistics and mathematics etc.

The first viral genome of bacteriophage φ174 is completed by Fred Sanger and his colleagues (Sanger et al., 1977), and the first complete eukaryotic genome is sequenced by Goffeau in 1996 (Goffeau et al., 1996). The subject for this project is a yeast call Saccharomyces cerevisiae. A lot of efforts from over 600 researchers in 100 laboratories are involved in order to obtain this genome. In the S. cerevisiae genome, there are about 13 Mb of DNA located in 16 different chromosomes. With the availability of the complete genome, Cherry et al. have unified the physical map with the genetic map (Cherry et al., 1997). The physical map can be obtained directly from DNA sequencing, while the genetic map by recombination analysis, which we will discuss in more details later.

The complete collection of DNA in Homo sapiens is called the human genome. The variations in the human genome can explain the differences between people, like the physical feature differences and the different disease states. The sequencing of the human genome has been achieved with the cooperation from the international community through the International Human Genome Sequencing Consortium

(IHGSC). On 15th February 2001, IHGSC reported the first draft version of the human genome (IHGSC, 2001). Nearly at the same time, Venter and colleagues (Venter et al., 2001) have reported their own Celera Genomics version of the draft sequence. As a brief summary of the sequencing results, it is estimated that there are about 30,000 to 40,000 genes in the human genome. And more than 98% of the genome is of the non-coding parts.

The fundamental unit for the human DNA is called the base. There are more than 6 billion of these chemical bases in the 23 pairs of chromosomes of the human genome. A specific position in the genome is called a locus (Sham, 1998). As said above, a genetic polymorphism refers to the existence of different DNA sequences at the same locus among a population. These different sequences are called alleles. In each base of the sequence, there can be any one of the four different chemical entities, which are adenine (A), cytosine (C), guanine (G) and thymine (T). Inside these genomic sequences, there contain the information about our physical traits, our resistance power to diseases and our responses to outside chemicals.

1.2.1 Single Nucleotide Polymorphisms (SNPs)

In most of the regions of any two human chromosomes, there exists identical sequencing. Nevertheless, there are regions of different sequencing. The differences in sequences can be grouped into large-scale chromosome abnormalities and small-scale mutations. The abnormalities include the loss or gain of chromosomes, and the breaking down and rejoining of chromatids. This can be found in tumor cells for example. The smaller-scale mutations can be further classified into: base substitutions, deletions or insertions (Taylor et al., 2005).

The most common type of genetic variations is that of differences in individual bases. They are called single nucleotide polymorphisms (SNPs, pronounced as "snips"). The HapMap project is to genotype the single nucleotide polymorphisms in the whole human genome (HapMap, 2005). The single nucleotide polymorphisms (SNP) is a common type of this small-scale mutation, and is estimated to occur once every 100–300 base pairs (bp) and the total number of SNPs identified reached more than 1.4 million. As an illustrative example, let's consider the five chromosome segments below:

1. ATCA**A**GC**C**A
2. ATCA**A**GC**A**A
3. ATCA**T**GC**C**A
4. ATCA**A**GC**C**A

We can see that in the fifth and eighth columns of the sequences, there exist some single nucleotide polymorphisms. These columns of SNPs are bold and the underlying minor variants are also underlined. For example, in the fifth base, sequences 1, 2 and 4 have the base of the adenine (A). But, in the sequence 3, the fifth base is a thymine (T) and is called a minor variant. Similarly, in the eighth base, sequences

1, 3 and 4 have the base of the cytosine (C), but that of the sequence 2 is of the adenine (A).

In the above example, we can also decide the tag SNPs that are needed. For the first, second, third, forth, sixth, seventh, and ninth bases, the corresponding building units are the same for each of the four sequences. For example, in the first base, all of them are the adenine (A). They are just the ordinary segments without any observable mutations. As said, the fifth and the ninth bases are of SNPs. In the fifth base, the minor variant (a T) occurs in the sequence 3, and in the ninth base, the minor variant (an A) occurs in the sequence 2. We can see that the distributions of the minor variants in these two bases are not similar to each other, so all these two SNPs are needed to be genotyped for medical analysis. Now, assume that we also have the tenth base, as followed:

1. ATCAAGCCAA
2. ATCAAGCAAT
3. ATCATGCCAA
4. ATCAAGCCAA

We can observe that the tenth base is a SNP and the minor variant occurs on the chromosome 2. The distribution of the minor variant of the eighth base and the tenth base are the same. Thus, one SNP (either the eighth base or the tenth base) is enough for representing these two SNPs. For example, assume that a disease T is caused by the minor variant T in the tenth base. Thus, sequence 2 will carry the disease T. If we know the disease distributions among the sequences instead and would like to know the potential variants, then, the results of genotyping the eighth base and the tenth base are the same. If we genotype the eighth base, we can see that the minor variant A occurs in the sequence 2 and that it is the sequence 2 that has the disease T. So we can identify that any member within the group (the eighth base or the tenth base) can be the variant for the disease. If we genotype the tenth base, we can have the same observations. In this example, we can see that we can save the genotyping cost by one-third, while we can still get the same association results as that of genotyping all SNPs.

1.2.2 Disease Studies with SNPs

As many common diseases are influenced by multiple genes and other environmental factors as well, it is not easy to assess their overall effect on the disease process. The genetic predisposition refers to a person's potential to develop a disease based on genetic and hereditary factors. The genetic factors can affect the susceptibility of a person to the disease, and may also influence the patient's response to drug therapy. The study of the SNPs can be helpful for the medical scientists to estimate the patients' responses to drugs. Because some SNPs are usually located near the genes associated with the disease, they can serve as biological markers for pinpointing the disease on the human genome. The SNPs become helpful for the scientists during the screening process for locating the relevant genes associated with the disease.

Briefly speaking, when a researcher is going to screen the genes associated with the disease, DNA samples from two groups of individuals are collected and compared. One group is of the individuals affected by the disease, while another group is of unaffected individuals. The differences between the SNP patterns of these two groups are compared. The results can indicate the patterns that are highly likely associated with the disease-causing gene. The goal is to establish SNP profiles that are characteristic of the disease. These studies are called association studies.

This type of research study is a very active area, and there have been a lot of research reports about the application of SNP techniques for a variety of diseases. For example, Langers et al. (2008) evaluated the prognostic significance of SNPs and tumour protein levels of MMP-2 and MMP-9 in 215 colorectal cancer patients. Fisher et al. (2008) conducted a nonsynonymous SNP scan for ulcerative colitics in study of Crohn's disease, and identified a previously unknown susceptibility locus at ECM1. Bodmer and Bonilla (2008) provided a historical overview of the search for genctic variants that influenced the susceptibility of an individual to a chronic disease. Chambers et al. (2008) carried out a genome-wide association study of more than 300,000 SNPs for insulin resistance and related phenotypes. It is found that common genetic variation near MC4R is associated with waist circumference and insulin resistance.

1.2.3 HapMap Project for Genomic Studies

The HapMap can be regarded as a catalog of common human genomic variants. It compares the genetic sequences among different individuals for locating chromosomal regions where genetic variants are shared. With the availability of this information freely, it will enable the researchers to figure out genes involved in diseases and to estimate individual responses to medications and environmental factors. By the end of February 2005, 7 month ahead of the target date, the group completed the first draft of the human haplotype map (HapMap). It consists of 1 million markers (SNPs) of genetic variations. On July 20, 2006, the HapMap project released its phase II dataset, which contains genotypes, frequencies and assays for bulk download. The data also includes genotypes from the Affymetrix 500k genotyping array. In the phase II, there existed more than 3 million non-redundant SNPs. The preliminary release of HapMap Phase 3, containing genotype and pedigree information for 11 populations (including individuals in the original four from earlier phases of the project), is available on May 27, 2008.

As the results from HapMap project have been becoming available to the researchers, the HapMap data have been applied in different genomic studies. For example, Cho (2008) used information on the correlation patterns observed from the HapMap databases to design genotyping platforms for the study of the inflammatory bowel disease. Hashibe et al. (2008) used the 163 SNPs genotyped by HapMap in the vicinity of the ADH gene cluster in the study of upper aerodigestive cancers. In the study of human bladder cancer, Majewski et al. (2008) employed the recombination rates based on HapMap and Perlegen83 data from UCSC and found seven HapMap recombination hotspots within the LOP peak. Gianotti et al. (2008) applied the phase II genotyping data from the HapMap project for the study of genetic variation on obesity and insulin resistance in male adults.

1.2.4 Potential Contributions of the HapMap Project to Genomic Analysis

It is expected that the HapMap project can provide a database that would be very useful for future studies on diseases. With the information of dense SNP genotyping in the second phase of the HapMap project, this can reduce many of the work and cost of the genomic searching of the disease genes as the genomic information like the tag SNPs of the genome is available. As the genotyping of the SNPs can be reduced to the set of tag SNPs, it is estimated that a saving of up to about 95% with the current brute force disease-searching approach can be achieved.

Another potential contribution of the HapMap project is that, with the information of the genomic variation, we can identify those variations that have an effect on good health. These variations may be the ones that can protect us from infectious diseases or that can enable the human being to live longer. Or, they may be the variations that can affect the individual's response to therapeutic drugs, toxic substances and environmental factors. With the availability of this information, it becomes possible to develop therapies and preventive strategies that are tailored to fit each individual's unique genetic characteristics. These customized medical treatments can maximize the effectiveness of the treatments and at the same time minimize the their side effects.

The knowledge from the HapMap can also be a guide for the association studies for the disease analysis. There is a hypothesis about the common-disease-common variance. It states that the risk of getting common diseases should be influenced by genetic variants that are also common in different populations. It is estimated that about 90% of sequence variation among individuals are caused by common variants (Kruglyak and Nickerson, 2001). It is also observed that, in most cases, each of these variants comes from single historical mutation event. Thus, they are associated with nearby variants that were presented on the ancestral chromosome where the mutation occurred. There is currently not enough data to assess this hypothesis generally, even though more and more widely distributed genetic variants are found to be associated with common diseases, such as diabetes, stroke and heart attacks. With the genotyping results from HapMap, it is expected that this can enable us to learn more about these links between the common disorders and our genes and genomic variations.

In the association studies, the traditional approach is to test each putative causal variant for identifying the correlation with the target disease. This is called the direct approach. The direct approach has the disadvantage of being expensive. One has to search the entire genome for any variants so that one can determine the disease associations. Thus, the scale of genotyping experiments required is very large and currently the approach is limited to the sequencing of the functional parts of candidate genes. With the HapMap information, an alternative approach is possible (Intl. HapMap Consortium, 2003). With this alternative approach, the sequencing costs would become much lower, as only a subset of the genomic variants serve as genetic markers for detecting association between a particular genomic region and the disease. The markers are not necessarily functional and the causative-variant search can be limited to the regions that have significant association with the disease.

Lastly, another potential contribution is that, in HapMap project, the population origins of the samples are kept and it becomes more efficient to analyses the population history and do inferences about the various degrees of relatedness of different populations. This population-history work can be helpful for biomedical researches. Nevertheless, issues like ethic issue may arise with this identification of population origins. Care has been taken to avoid the conflicts with the individual population customs or culture. For example, the American-Indian tribes have not been chosen because the findings may conflict with their religious and cultural understandings of their origins (Intl. HapMap Consortium, 2004).

1.2.5 Haplotypes, Haplotype Blocks and Medical Applications

Even though recombination events repeat generations after generations and segments of the ancestral chromosomes in a population are shuffled, there are still some segments that have not been broken up by recombination. These segments occur as regions of DNA sequences shared by multiple individuals, and are separated by places where recombination has occurred. These segments are call haplotypes. The haplotypes can enable the medical scientists in the search for genes in the diseases and in the study of important genetic traits.

Daly et al. (2001) began the studies of the haplotypes for the linkage disequilibrium (LD) analysis and compared these results with the results from single-marker LD. It is shown that the noises, which are presumably caused by the marker history etc., disappear when using the haplotype-based LD. Daly's results also show that there exists a picture of discrete haplotype blocks that are of order tens to hundreds of kilobases. Inside each block, there is only a little diversity, while between the blocks there are punctuations that show the potential sites of recombination. Daly et al. have observed that, over a long distance, most haplotypes can be cataloged into a few common haplotype categories. The idea of the haplotype blocks has come from studies like that of Gabriel et al. (2002). Gabriel et al. showed that the human genome can be divided into haplotype blocks, which are defined as regions of little historical recombination and of only a few common haplotypes.

Different studies have conducted the haplotype analysis for the disease studies. For example, Levy-Lahad et al. (1995) found that there was positive evidence for linkage with markers on the chromosome 1 for the Alzheimer's disease. Tishkoff et al. (2001) carried out haplotype analysis of A- and Med mutations at this locus for the study of malarial resistance. Singleton et al. (2003) discovered a chromosome

4p15 haplotype segregating with parkinsonism and essential tremor, with suggestive evidence for linkage to PARK4. Herbert et al. (2006) identified a core haplotype block containing rs7566605 in their association study of adult and childhood obesity. Couzin and Kaiser (2007) provided an overview of the application of genome-wide association study for common diseases like diabetes, heart disease, inflammatory bowel disease, macular degeneration, and cancer. The studies derive the power from Hapmap and Haplotype Map that catalogs human genetic variation.

1.2.6 Genomic Analysis with Microarray Experiments

Microarray is a solid substrate where the DNA is attached to in an ordered manner at high density (Geschwind and Gregg, 2002). Among the high-throughput methods of gene expression, the microarray has been the most widely used one for assessing the differences in mRNA abundance in the biological samples. With the work of Patrick Brown and his colleagues (DeRisi et al., 1996), microarray has been gaining its popularity.

In a single microarray experiment, the expression levels of as many as thousands of genes can be measured simultaneously. Thus, it can enable the genome-wide measurement of gene expression. This is a large improvement over the situation of "one gene per experiment" in the past.

The microarray technology can also enable us to have the gene expression values at different time points of a cell cycle. In the literature, different methods have been developed to analyze gene expression time series data, see for instance (Costa et al., 2002; Yoshioka and Ishii, 2002; Tabus and Astola, 2003; Syeda-Mahmood, 2003; Wu et al., 2003). The construction of genetic network from gene expression time series is tackled in (Kesseli et al., 2004; Tabus et al., 2004; Sakamoto and Iba, 2001). The visualizing of the gene expression time series is discussed in studies (Zhang et al., 2003; Craig et al., 2002). More details about the microarray technology are available in the Section 3.4.

1.3 Case Studies: Building Data Mining Algorithms for Genomic Applications

With the advances in the technology for the genomic analysis, it is not unusual that millions of data records are produced and needed for investigation in one genomic study (Bergeron, 2003). It becomes very costly to search for any meaning information from these datasets by human inspection. Advances in the improvement and new designs of data mining algorithms are needed for modeling genomic problems efficiently. In these cases, data mining algorithms are very useful to extract the patterns from the data and to present it in such a way that can enable us to have a better understanding of the structure, relation, or function of the subjects.

In order to illustrate the development process of tailor-made data mining algorithms for the genomic analysis, three case studies are highlighted to show the motivations, the algorithms, the computational considerations and the performance evaluation. In the first case study, we have developed clustering and graph algorithms for the problem of tag-SNP selection, which can combine functional and linkage disequilibrium information. It has been shown to reduce efficiently the costs of genotyping. In the second case study, non-parametric method of constrained unidimensional scaling has been proposed for constructing linkage disequilibrium map (LD map), which may have the medical potentials of locating disease genes etc. Thirdly, hybrid algorithms of principal component and neural network have been developed for the continuous microarray time series, which have been shown to have better predictability than the other methods and which offer us an efficient tool for investigating continuous microarray time series.

1.3.1 Building Data Mining Algorithms for Tag-SNP Selection Problems

With the results from the genomic projects like the HapMap Project, it is estimated that there exist about 10 million single nucleotide polymorphisms (SNPs) in the human genome. Although only a proportion of these SNPs are functional, all can be used as markers for indirect association studies to detect disease-related genetic variants. With such a large number of SNPs, the complete screening of all the SNPs in a genomic region becomes an expensive undertaking. It is much more cost-effective to develop tools for selecting a subset of informative SNPs, called tag SNPs, in the medical or biological analysis (Johnson et al., 2001).

We have formulated this problem of selecting tag SNP as a hierarchical clustering problem and developed a suitable similarity function for measuring the distances between the clusters (Ao et al., 2005; Ng et al., 2006; Sham and Ao et al., 2007). Hierarchical clustering algorithms can be classified into two types, the agglomerative algorithms and divisive algorithms, according to their procedures of grouping or dividing the data points. In the agglomerative algorithms, they produce a sequence of clustering with decreasing number of clusters m at each step. On the other hand, divisive algorithms give us a clustering sequence of increasing number of clusters at each step. The final product is a hierarchy of clustering with these algorithms. In our works, we have applied the agglomerative algorithms for the tag SNP selection problem. Therefore, we shall restrict our discussion to the agglomerative algorithms. For their computational requirements, Murtagh (1983, 1984 and 1985) has discussed about the implementations for widely used agglomerative algorithms and the computational time complexity is of $O(N^2)$.

1.3.2 Building Algorithms for the Problems of Construction of Non-parametric Linkage Disequilibrium Maps

There are several different advantages of the linkage disequilibrium maps (LD maps) for human genome. The LD map can provide us with a much higher resolution of the biological samples than the traditional linkage maps. The other advantages of LD maps are the revealing of the recombination patterns, the facilitating of the optimal SNP/marker spacing, and the increasing of the power for localizing disease genes etc.

The first LD maps were proposed by Maniatis and colleagues (2002) and are based on the Malecot equation. The derivation of this LD map is parametric and requires the estimation of three coefficient parameters. Nevertheless, these estimated parameters are found to have large variances among different populations.

We have formulated this LD mapping problem as a constrained unidimensional scaling problem (Ao et al., 2005, 2007; Ao, 2008). Our method, which is directly based on the measurement of LD among SNPs, is non-parametric. Therefore its underlying theory is different from LD maps derived from the given Malecot model. For solving this constrained unidimensional scaling problem, we have formulated it as a quadratic optimization problem. Different from the classical metric unidimensional scaling problem, the constrained problem is not an NP-hard combinatorial problem. The optimal solution is determined by using the quadratic programming solver.

1.3.3 Building Hybrid Models for the Gene Regulatory Networks from Microarray Experiments

The neural network is one of the machine learning tools that can reduce noises and make prediction reliably. A key property of the neural network is its ability of learning for further improving its performance (Huang et al., 2004). The learning process starts with the stimulation by the environment. Then, the neural network will have changes in its structure and parameters as a result of this stimulation. These changes will bring the network improvement in its response to the environment. In the learning process, there can be different objective tasks to achieve, like the function approximation, control, pattern recognition, filtering and prediction etc. We have employed the neural network for the function approximation and prediction of the cell cycles time series microarray data.

Different genetics studies have successfully employed the PCA-NN as a classifier of gene types, with continuous inputs and discrete outputs. In this work, we have been developing an algorithm for testing the predictability of the gene expression time series with the PCA and NN components on a continuous numerical inputs and outputs basis. The contribution of our work lies in the fact that we have been developing a more realistic model for the gene network from a continuous prospective

(Ao et al., 2004; Ao and Ng, 2006). A microarray dataset can be considered as a matrix of gene expression values at various conditions. Each entry in the matrix is a numerical number called expression value. The algorithm can fully utilize the information contained in the gene expression datasets. It can be considered as an extension of the linear network inference modeling, while previous models have often needed the linearity assumption or employed discrete values instead.

The formulation of our PCA-NN algorithm is quite computationally efficient. The input vectors for the time series analysis are the expression levels of the time points in the previous stages of the genes' life cycle. These input vectors are processed by the PCA component. Then, we use these post-processed vectors to feed the neural network predictors. In order to avoid over-training of the network, we have adopted the AIC test and cross-validation to study the optimal setting of the neural network structures and the network's stability. The AIC test can restrict the number of parameters of the network and thus can increase the computational performance.

The possibility of adding the GA component will be explored too. We can set the inclusion or exclusion of each gene in the building of the gene expression network for a particular gene. This can simplify the gene network. It has been shown to be able to reduce the computational complexity of the originally NP-hard gene expression analysis efficiently, as pointed out by Keedwell in his work on genetic algorithm.

Chapter 2
Data Mining Algorithms

In this Chapter, we will describe about several groups of the basic data mining algorithms. In the first section of dimension reduction and transformation algorithms, we will talk about the feature selection, feature extraction methods like principal component analysis and independent component analysis etc. In the section of machine learning algorithms, topics like logistic regression models, neural network models, fuzzy systems, ensemble methods, support vector machines and hybrid intelligent techniques etc. will be covered. Then, we will also talk about the clustering algorithms like hierarchical clustering, partition clustering spectral clustering, and their considerations. In the section of graph algorithms, topics like computer representations of graphs, breadth-first search algorithms, and depth-first search algorithms will be covered. This Chapter will conclude with the discussion of several popular numerical optimization algorithms like steepest descent method, Newton's method, sequential unconstrained minimization, reduced gradient methods, and interior-point methods.

2.1 Dimension Reduction and Transformation Algorithms

Before extracting meaningful patterns from the data, there are sometimes the needs for the transformation and reduction of the data. These needs may arise due to the fact that the original dataset is too large in dimension. Both reduction and transformation can support the data-mining process when used properly. The datasets may be reduced to the minimum possible size by tactics like sampling or summary statistics etc., while still satisfying our analysis requirement. The transformation can be achieved by translating one type of data to another through mathematical operations or mappings. The noise level in the data may be reduced by the eliminating irrelevant components with suitable transformation. Transformation tools like principal component analysis (PCA) and independent component analysis (ICA) can be employed to find out the dominant components of the dataset. We have applied these algorithms for the microarray time series data, and we will compare their respective performances in our hybrid models.

Sio-long Ao, *Data Mining and Applications in Genomics*,

2.1.1 Feature Selection

Feature selection is the technique to select a subset of relevant features for constructing robust learning models. The optimal feature selection for supervised learning problems requires an exhaustive search of all possible subsets of features. This is not practical for large data sets. Greedy algorithms have been developed to overcome this drawback. Generally speaking, feature selection has the advantages like addressing the curse of dimensionality, speeding up data mining process and enhancing generalization capability. It can improve the model interpretability by providing a better understanding of the important features (Liu and Hiroshi, 1998). Feature selection techniques have been employed in many bioinformatics applications. In the domain of gene selection, this method is also called discriminative gene selection, and it can select the influential genes based on data sets genotyped from microarray experiments etc. Feature selection techniques like filter methods, wrapper methods and embedded method has been applied in the microarray domain. There are also applications for the content analysis and signal analysis in sequence analysis. Applications of the feature selection for mass spectrometry technology, which is a new and attractive framework for disease diagnosis and protein-based biomarker profiling, have also been reported. Other applications include, for example, SNP analysis, and text and literature mining etc. Saeys et al. (2007) provided a comprehensive review of the feature selection techniques in bioinformatics.

2.1.1.1 Information Gain

Information Gain is a measure for deciding the relevance of an attribute in the feature selection (Mitchell, 1997). In data mining applications, it is used to define a preferred sequence of attributes to narrow down the state of a random variable *X*. Generally speaking, an attribute of high information gain is selected over other attributes of low information gain.

The expected value of the information gain is the reduction in the entropy of *X* with the knowledge of the state of the random variable *A*. Mathematically, it is given as followed:

$$IG(Ex, a) = H(Ex) - H(Ex \mid a)$$

where *Ex* is the set of all training examples, and *H* is the entropy function. Entropy is a measure of the uncertainty associated with a random variable. For a discrete random variable X of possible values $\{x_1 \ldots x_n\}$, $H(X) = -\sum_{i=1}^{n} p(x_i) \log_2 p(x_i)$, where $p(x_i) = \Pr(X = x_i)$ is the probability mass function of X.

2.1.1.2 Chi-Square Distribution

Chi-square distribution is commonly used in inferential statistics, for example, in the statistical significant tests. It has been proven that the chi-square distributions can be used to approximate other distributions under some assumptions if the null hypothesis is true. Chi-square distribution can be used for the test of the independence of two criteria of classification of qualitative data (Johnson et al., 1994). Let X_i be k independent, normally distributed random variables with mean 0 and variance 1. The distribution of the random variable

$$Q = \sum_{i=1}^{k} X_i^2$$

is of the chi-square distribution, and is usually written as $Q \sim \chi_k^2$.

2.1.1.3 Recursive Feature Elimination

Recursive feature elimination can be useful when removing several features at a time. It can obtain a small feature subset, and can be represented in three iterative procedures: (1) training the classifier; (2) computing the ranking criterion for all features; (3) removing the feature with smallest ranking criterion. For computational considerations, it may improve the efficiency by removing several features at a time, even though it may degrade the classification performance to a certain extend. In such a case, the method produces a feature subset ranking. For cases when the features are removed one at a time, the method produces a corresponding feature ranking. This recursive feature elimination method can apply in the genomics studies, for example, in the gene selection for cancer classification (Guyon et al., 2002).

2.1.2 Feature Extraction

Feature extraction is to transform the input data into a reduced representation set of features (feature vectors). Its main difference with the feature selection techniques is the change of the original representation of the variables. Feature extraction can be very usefully when the input data is large and when there is redundant information in the data. The feature extraction algorithms can return the feature vectors with the relevant information for performing the target task instead of the full size input. They can improve the computational performance as well as address the issues like over-fitting and noises in the experiments.

2.1.2.1 Principal Component Analysis

Among the tools of the dimension reduction and transformation, the principal component analysis (PCA) is a popular tool for many researchers. Its basic idea is to find the directions in the multidimensional vector space that contribute most to the variability of the data. The principal component analysis was applied to reduce the dimensionality of the gene expression data in studies (Hornquist et al., 2003; Bicciato et al., 2003; Taylor et al., 2002; Yeung and Ruzzo, 2001, etc.). The focuses are on the effective dimensional reduction by the PCA, the analysis of the compressed space and the assistance of the PCA for the classification and the clustering. Khan et al. (2001) applied the PCA and neural network for the classification of cancers using gene expression profiling. More details of the principal component analysis are available in Chapter 6.

2.1.2.2 Multifactor Dimensionality Reduction

Multifactor dimensionality reduction (MDR) is a constructive induction algorithm that converts variables (or attributes) to a single attribute. It can identify the interactions among discrete variables that influence a binary outcome. The technique has been widely applied in bioinformatics problems. MDR can detect the gene-gene interactions in the genetic studies of common diseases such as bladder cancer, breast cancer, cardiovascular, schizophrenia, and type II diabetes. For example, MDR was applied to reveal high-order interactions among estrogen-metabolism genes in sporadic breast cancer (Ritchie et al., 2001). Cho et al. (2004) applied the technique for Type 2 diabetes. Coffey et al. (2004) employed the method to detect gene-gene interactions on risk of myocardial infarction.

2.1.2.3 Nonlinear Dimensionality Reduction

Nonlinear dimensionality reduction, also called manifold learning, is a new method to reduce the data dimensionality in a nonlinear way. It can provide a way to understand and visualize the structure of complex data sets. It assumes that the high-dimensional data lies on an embedded non-linear manifold within the higher dimensional space. If the manifold has a low enough dimension, the data can be visualized in this low dimensional space as well. Different methods (for example, Lee and Verleysen, 2007) were available in the domain of the estimation of the intrinsic dimension (PCA estimator, Local PCA estimator, and correlation dimension etc.), distance preservation (multidimensional scaling, Sammon's nonlinear mapping, and curvilinear component analysis, etc.), and topology preservation (self-organizing maps, generative topographic mapping, and locally linear embedding, etc.). Nonlinear dimensionality reduction has been applied in different computational biology problems. For example, Wang et al. (2008) has applied the technique to improved classifier for computer-aided polyp detection in CT colonography. Georgiadis et al. (2008) applied the technique for discriminating between metastatic and primary brain tumors (gliomas and meningiomas) on MRI, employing textural features from routinely taken T1 post-contrast images.

2.1.2.4 Kernel PCA

Kernel principal component analysis (kernel PCA) is an extension of principal component analysis (PCA). Kernel PCA uses the kernel methods, that is, the originally linear operations of PCA are implemented in a reproducing kernel Hilbert space with a non-linear mapping. Thirion and Faugeras (2003) built a redundant representation of the data through the nonlinearity of the kernel PCA for the analysis of fMRI data. The model was useful in the characterization of subtle variations in the response to different experimental conditions. Tome et al. (2007) proposed the kernel PCA for the correction of univariate, single channel EEGs. The proposed kernel method employed a greedy approach to use a reduced data set to compute a new basis onto which to project the mapped data in feature space. Their results show good performance in removing artifacts like eye or head movements.

2.1.2.5 Latent Semantic Analysis

Latent semantic analysis (LSA) studies the relationships between a set of documents and the term they contain (Deerwester et al., 1990). LSA use a term-document matrix to described the occurrences of terms in documents, and produces a set of concepts related to the documents and the terms. Vanteru et al. (2008) applied the LSA method to link the PubMed to the Gene Ontology for ontology-based browsing. The PubMed is the current most widely used repository for bio-literature, and it consists of about 17 million abstracts as of 2007, requiring methods for efficiently retrieving and browsing. The results showed that the proposed LSA technique outperformed the string comparison based techniques in associating the relevant abstracts to the GO terms. Ganapathiraju et al. (2008) proposed a LSA method for the prediction of transmembrane (TM) helices with high accuracy. The proposed method can extract features from protein sequence and have the potential for applications in other sequence-based analysis problems as well.

2.1.2.6 Independent Component Analysis

The independent component analysis (ICA) is a recently developed theory (Hyvärien et al., 2001; Comon, 1994 and Jutten and Herault, 1991). Its objective is to make the transformed entries mutually independent (Theodoridis and Koutroumbas, 2003). Biswas et al. (2008) applied Independent Component Analysis to gene expression traits derived from a cross between two strains of Saccharomyces cerevisiae. It showed that the dimension reduction method is a useful approach for probing the genetic architecture of gene expression variation. Esposito et al. (2008) proposed different strategies for combing the ICA results from individual-level and population-level analyses of brain function to study of the effect of aging on the DM component. Liu and Huang (2008) showed that ICA can further improve the performance of rotation forest in cancer classification with the microarray data sets. More detailed description of the independent component analysis is available in Chapter 6.

2.1.3 Dimension Reduction and Transformation Software

Saeys et al. (2007) provided an updated survey of the software for feature selection. General purposed FS software includes, for example, WEKA (http://www.cs.waikato.ac.nz/ml/weka),

MLC + + (http://www.sgi.com/tech/mlc), Spider (http://www.kyb.tuebingen.mpg.de/bs/people/spider), SVM and Kernel Methods Matlab Toolbox (http://asi.insa-rouen.fr/~arakotom/toolbox/index). Microarray analysis FS software includes software like SAM (http://www-stat.stanford.edu/~tibs/SAM/), GA_KNN (http://dir.niehs.nih.gov/microarray/datamining/), GenePattern (http://www.broad.mit.edu/genepattern). SNP analysis FS software include CHOISS (http://biochem.kaist.ac.kr/choiss.htm), MLR-tagging (http://alla.cs.gsu.ed/~software/tagging/tagging.html), WCLUSTAG (http://bioinfo.hku.hk/wclustag).

The principal component analysis (PCA) is available in popular data analysis software like Matlab. The website (http://sourceforge.net/projects/mdr/) has the open-source multifactor dimensionality reduction software package. It is developed by the Computational Genetics Laboratory at the Norris-Cotton Cancer Center and Dartmouth Medical School in Lebanon, New Hampshire. The SVM and Kernel Methods Matlab Toolbox is available for download at http://asi.insa-rouen.fr/enseignants/~arakotom/toolbox/index.html. The Semantic Indexing Project is to create tools to identify the latent knowledge found in text, and the source code is available for download at: http://knowledgesearch.org/. A Matlab implementation of the ICA is available at: http://www.cs.helsinki.fi/u/ahyvarin/whatisica.shtml.

2.2 Machine Learning Algorithms

The machine learning algorithms have often been employed for pattern matching and pattern discovery. Machine learning is itself a collection of methods that result from the convergence of several disciplines like statistics, biological modeling, adaptive control theory and artificial intelligence (AI). Its spectrum of tools is wide and includes inductive logic programming, genetic algorithms, neural network, Bayesian networks, and hidden Markov Models, etc. Regardless of the divergences of the underlying technology, we can see that they usually have the following steps (Bergeron, 2003). First, the input data are fed into a comparison engine that compares the data with assumed model. Then, the comparison results are used for initializing some changes to the data or some modifications to the assumed model. And the evaluation engine gives us the performance results based on the modified model. These results are checked against our pre-defined criterion. If the criteria are not met, the above steps are repeated until the stopping criteria are satisfied. We will talk about logistics regression models, decision tree algorithms and neural networks in this section.

2.2.1 *Logistic Regression Models*

In genomic study, it is interesting to study the effects of some variables on the target variable, for example, the effects of some genes on certain diseases. One sensible way for expressing this relationship between the variables is by some sort of mathematical equation. When the variables are continuous and the relationship is believed to be linear, a suitable mathematical equation will be the simple linear regression model:

$$y_i = \sum_{j=0}^{k} \beta_j x_{ji} + \varepsilon_i$$

where there are k input variables $x_1, x_2, x_3, \ldots, x_k$ with unknown parameters $\beta_0, \beta_1, \ldots, \beta_k$, and the term ε_i is the unobservable random error representing the residual variation (Shoukri and Pause, 1998). The methods of the least squares and the maximum likelihood are commonly used for the parameter estimation of the above regression model.

For the cases of binary target variable y, the direct application of the above regression is not well desired, as some of the assumptions for this model are not satisfied. Then, for this binary target problem, we need to apply the logistic transformation, which is fundamental for the logistic regression models. Let's consider the case that binary variable y refers to the disease status, with the value 1 for getting the disease and 0 for no disease in the sample. And, the variable X is a risk factor for the disease, with normal distribution. Let $P(D) = P[y = 1] = \pi$ and $P(\bar{D}) = \Pr[y = 0] = 1 - \pi$. Let $N(\mu_1, \sigma^2)$ and $N(\mu_2, \sigma^2)$ be the conditional distributions of x for the diseased population and for the non-diseased population respectively. Let $p = \Pr[y = 1 \mid X = x] = p(y = 1, X = x)/f(x)$, where

$$f(X \mid D) = 1/\sigma\sqrt{2\pi}\exp\left[-(x-u_1)^2/2\sigma^2\right]$$

$$f(X \mid \bar{D}) = 1/\sigma\sqrt{2\pi}\exp\left[-(x-u_2)^2/2\sigma^2\right].$$

With Bayes' theorem,

$$p = \frac{f(X \mid D)p(D)}{f(X \mid D)p(D) + f(X \mid \bar{D})P(\bar{D})}$$

After some manipulation, we can have

$$p = \frac{\exp[\beta_0 + \beta_1 x]}{1 + \exp[\beta_0 + \beta_1 x]}$$

where

$$\beta_0 = -\ln\frac{1-\pi}{\pi} - \frac{1}{2\sigma^2}(\mu_1 - \mu_2)(\mu_1 + \mu_2)$$

and

$$\beta_1 = \frac{\mu_1 - \mu_2}{\sigma^2}$$

Then, we can have the log-odds result

$$\ln\left(\frac{p}{1-p}\right) = \beta_0 + \beta_1 x$$

The above log-odds is a linear function of the input variable X, and this logarithmic transformation on the odds is called "logit". The maximum likelihood method can be used to estimate the parameters of this model. Pearson's chi-square and the likelihood ratio test can be used for measuring the fitness between the responses of the postulated model and the observed data.

2.2.2 *Decision Tree Algorithms*

Decision tree algorithms are decided with a sequence of decision rules for splitting a large set of records into successively smaller sets of records. The objective is to obtain more homogeneous smaller sets with respect to a particular target variable. These algorithms have been found powerful and popular for both classification and prediction problems (Berry and Linoff, 2004). One of the reasons for its popularity is that the decision rules can also be expressed in English for easy understanding and presentation. The following figure (Fig. 2.1) illustrates what the graphic output of the decision tree looks like with the software SAS.

Even though there exist different decision tree algorithms, all of them share the same basic procedure. That is to repeatedly split the data records into smaller groups, so that, in each new generation, the new sub-groups will have greater purity

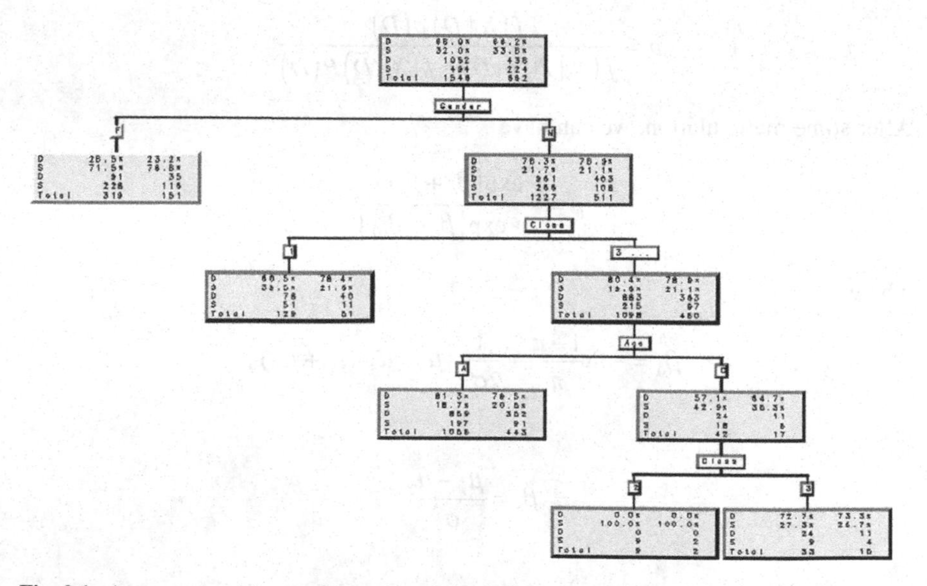

Fig. 2.1 An example of the graphic output of the decision tree with the software SAS

than its ancestors with respect to the target variable. The purity is a measure used to evaluation the performance of the potential splits. For categorical target variable, tests such as Gini, information gain, or chi-square are suitable for measuring the purity levels of the splits. For continuous numeric variable, tests like variance reduction and F-test are appropriate.

At the start of the decision tree building, the algorithms try to identify the input variable that can have the best performance of splitting the data with respect to the above purity. Then, once again, all the input variables are considered as the candidate splitters in the coming rounds. At the succeeding level of the tree, these subsets are further split according to the rule that works best for them. This process is exhaustive, as the algorithms proceed by checking each input variable in turn for building the splits.

The performance of a decision tree algorithm can be assessed with the test data, which consists of data records that have not been used for building the tree. The misclassification rate is popular for this assessment. The method of pruning can be applied to choose the tree with the minimal misclassification rate. The following figure (Fig. 2.2) illustrates this determination of the tree structure.

2.2.3 *Inductive-Based Learning*

Inductive learning is the process of learning by example, where the system tries to induce a general rule from the set of observed instances (Quinlan, 1990). Inductive methods can

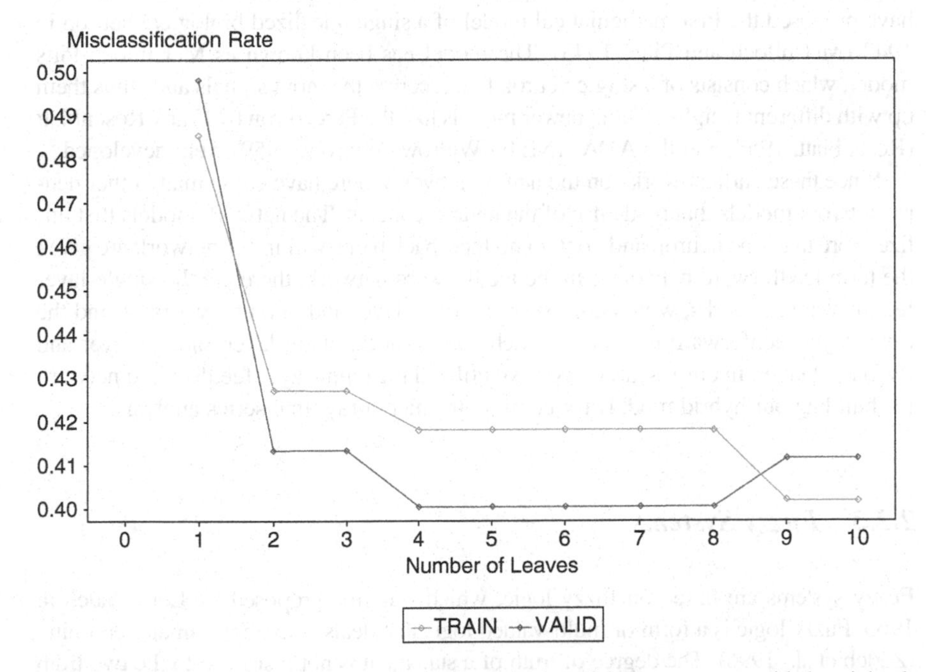

Fig. 2.2 An example for the illustration of the determination of the tree structure with the minimal misclassification rate

be characterized as search methods over a hypothesis spaces. To constraint the hypothesis space, relational learners introduce a partial order between hypotheses. Jia and Kitchen (2000) applied the inductive-based learning algorithm to predict the class distribution of object-contour segments in image similarity computation. Garcia-Gomez et al. (2005) proposed a hybrid method of stochastic context-free grammars (SCFG) and Hidden Markov Models (HMM) for modeling tRNA secondary structures. Given annotated public databases, the HMM and SCFG models are learned by means of automatic inductive learning methods. Significant results were obtained in the performed experiments on the tRNA sequence corpus and the non-tRNA sequence corpus.

2.2.4 Neural Network Models

Among the different tasks that machine learning tools can handle, one popular task is to filter the noises of the source data and then to made prediction basing on the extracted patterns of the source data. The neural network has been found to perform this filtering and prediction capability well. The network can first extract the vital signals and information from the source data. Then it can predict what the future signals and information will be, based on some function approximation. The filtering and prediction capability of the neural network have enabled it to become a popular advance tool for the time series prediction, e.g., the financial prediction of the future index, stock share prices, currency rates, the weather forecast, the traffic control forecast and the medical analysis.

The study of the neural network began after the Warren McCulloch and Walter Pitts have proposed the first mathematical model of a single idealized biological neuron in 1943 (McCulloch and Pitts, 1943). The model has been known as McCulloch-Pitts model, which consists of a single neuron that receives the input signals and sums them up with different weights. Then, newer models like the Perceptron by Frank Rosenblatt (Resenblatt, 1958) and the ADALINE by Widrow (Widrow, 1959) were developed.

Since these earliest works on the neural network, there have come many other neural network models that made use of the neuron concept. The network models that utilize more than one neuron and contain no feed-back paths within the network are given the term feedforward networks. In the feedforward network, there are the single-layer feedforward networks, which consist of the input layer and output layer only, and the multi-layer feedforward networks, which consist of the input layer, hidden layer and the output layer. In our research, we have utilized the multi-layer feedforward network for building our hybrid models for continuous microarray time series analysis.

2.2.5 Fuzzy Systems

Fuzzy systems are based on fuzzy logic, which was first proposed by Lotfi Zadeh in 1965. Fuzzy logic is a form of multi-valued logic that deals with approximate reasoning (Zadeh et al., 1996). The degree of truth of a statement is not restricted to the two truth values {true, false}. Instead, it can range between 0 and 1.For example, in a 1,000-ml bottle there are 200 ml of coke, one might define the bottle as being 0.2 full and 0.8 empty, for the two fuzzy sets, Full and Empty. The fuzzy set theory defines the fuzzy operators on these fuzzy sets. A difficulty with the fuzzy systems is that the appropriate Fuzzy Operators may not be known in advance. Ghazavi and Kim et al. (2006) proposed the fuzzy partitional clustering method known as Fuzzy C-Means (FCM) to overcome the limitations of hard clustering for the gene expression microarray data. Ghazavi and Liao (2008) proposed three fuzzy modeling methods including the fuzzy k-nearest neighbor algorithm, a fuzzy clustering-based modeling, and the adaptive network-based fuzzy inference system for medical data mining. The proposed methods were applied to the medical data of the Wisconsin breast cancer dataset and the Pima Indians diabetes dataset. Akdemir (2008) proposed a new method based on combining principal component analysis (PCA) and adaptive network-based fuzzy inference system (ANFIS) to diagnose the optic nerve disease from visual-evoked potential (VEP) signals.

2.2.6 Evolutionary Computing

Evolutionary computation techniques are often inspired by the biological systems of evolution (Jong, 2006). Evolutionary computing involves the iterative steps of the growth or development in a population. In the iterative process, the population is selected in a random search to achieve the target goal. It includes the metaheuristic optimization algorithms like genetic algorithms, evolutionary programming, evolution strategy, genetic programming, ant colony optimization and particle swarm optimization. Lamers et al. (2008) proposed a hybrid system of artificial neural networks trained via evolutionary computation for predicting viral co-receptor usage. The results show identification of R5X4 viruses with predictive accuracy of 75.5%. Ritchie et al. (2007) applied the multifactor dimensionality reduction (MDR) and grammatical evolution neural networks (GENN) to three data sets from GAW15 for the analysis of rheumatoid arthritis. Rowland (2003) proposed an approach to model selection in supervised learning with evolutionary computation with applications to metabolite determination and to disease prediction from gene expression data. To improve the diagnosis, prevention, and treatment of common, complex human diseases, Moore et al. developed a hierarchical dynamic systems approach based on Petri nets for generating biochemical network models that are consistent with genetic models of disease susceptibility. An evolutionary computation approach – grammatical evolution, is used as a search strategy for optimal Petri net models (Moore et al., 2005).

2.2.7 Computational Learning Theory

Computational learning theory examines formal models of induction to discover the common methods underlying efficient learning algorithms, and to identify the computational impediments to learning (Kearns and Vazirani, 1994). The emphasis is on rigorous mathematical analysis. It includes algorithms that can make predictions about the future based on past observations, algorithms that can learn from a teacher, and algorithms that can learn by interacting with the world around them. Here, a computation is said to be feasible if it can be completed in polynomial time. Positive results refer to a certain class of functions that can be learned in polynomial time. Negative results refer to a certain class that can not be learned in polynomial time. There exist different approaches to computational learning theory, like probably approximately correct learning (PAC learning), Vapnik-Chervonenkis (VC) theory, Bayesian inference, and algorithmic learning theory (Angluin, 1992). Based on the theory, practical algorithms like boosting (inspired by PAC theory), support vector machines (inspired by VC theory), and belief networks (inspired by Bayesian inference). Trumbower et al. (2006) applied the PAC learning model to classify static offline muscle strength observations with online rider performances. Miller and Eisenberg (2008) proposed a Bayesian inference procedure to identify residue pairs that are spatially proximal in a protein structure. The approach takes as input a multiple sequence alignment, and outputs an accurate posterior probability of proximity for each residue pair. Sajda (2006) provided a comprehensive survey of recent developments in machine learning, focusing on supervised and unsupervised linear methods and Bayesian inference, which had significant impacts in the detection and diagnosis of disease in biomedicine.

2.2.8 Ensemble Methods

Ensemble methods have been developed for improving the predictive performance of a given individual statistical learning algorithm. Ensemble methods can be applied to numerical prediction and classification problems. Previous studies have shown that ensemble methods can often provide more accurate prediction than any of the individual classifiers in the ensemble (Opitz and Maclin, 1999). The resulting classifier of the ensemble is generally more accurate than any of the individual classifiers making up the ensemble. Both the theoretical and empirical research has shown that the individual classifiers of a good ensemble need themselves to be accurate and to make their errors on different parts of the input space. Studies like Maqsood et al. (2004) show that ensemble methods can improve numerical prediction above that of the individual predictors.

A basic approach of the ensemble methods is to formulate a linear combination of some individual learning algorithms, instead of using one single fit of one algorithm. An estimation of a real-value function can be expressed mathematically as $g : R^d \to R$ with a d-dimensional predictor variable $\boldsymbol{X}$ and a 1-dimensional response/target $\boldsymbol{Y}$. A base procedure is a specific algorithm which yields one estimated function $\hat{g}(\cdot)$. It is possible to run different base procedures many times to have different estimated functions. An ensemble-based function estimates $\hat{g}_{en}(\cdot)$ by having a linear combination of the individual function estimates $\hat{g}_k(\cdot)$:

$$\hat{g}_{en}(\cdot) = \sum_{k=1}^{M} c_k \hat{g}_k(\cdot)$$

where c_k are the linear combination coefficients. The coefficients can simply assume averaging weights, or can assume different numerical values.

In an ensemble of neural networks, a collection of a finite number of neural networks is applied simultaneously for the same task. This ensemble approach originated from the work of Hansen and Salamon (1990), which have applied many neural networks and combined their predictions to form the ensemble. The results showed that significant improvement can be obtained for the generalization ability of a neural network system. Previous studies have applied the ensemble method for the classification problems of microarray data. The ensemble neural networks with combinational feature selection have been applied to the microarray experiments for the tumor classification and remarkably improved results have been obtained (Liu et al., 2004). It has been shown that the ensemble method is able to reduce the instability of the individual neural networks and help alleviate the problem of trapping the neural network into local minima. The ensemble method can cancel the noise part among its individual networks and retain the fitting to the regularities of the data.

2.2.9 Support Vector Machines

Support vector machines (SVMs) were developed at AT&T Bell Laboratories by Vapnik and his co-workers (Boser et al., 1992). Because of its industrial context, the support vector approach had a sound orientation towards real-world applications (Smola and Scholkopf, 2004). Support vector machines can be divided into two main types: support vector machines for classification (SVC) and support vector machines for regression, in short, support vector regression (SVR). Support vector machines have many mathematical features attractive for gene expression analysis (Brown et al., 2000). These include the sparseness of solution for large data sets, the flexible capability for handling large feature spaces, the robust identification of outliers, etc. SVC has found applications in different bioinformatics domains, for example, protein secondary structure prediction (Guo et al., 2004), cancer classification (Guyon et al., 2002), and enzyme family classification (Cai et al. 2004). While SVR has been applied to the time series prediction applications and excellent performances were obtained (Muller et al., 1997) and (Mattera and Haykin, 1999), only until recently, the SVR has been applied for the bioinformatics problems like the missing value estimation (Wang et al., 2006) and the smoothing of expression profiles (Martin et al., 2007) for DNA microarray gene expression data.

Support vector machines are based on the structural risk minimization principle from statistical learning theory. The structural risk minimization principle can minimize both empirical risk and confidence principle and avoid the "over-fitting" problem. The basic principle of the support vector machines is to map the samples from the low-dimension input space into a much higher dimensional space with a kernel function. Then, quadratic programming is applied to search the global optimal solution to the corresponding problem. Support vector regression constructs the fitted regression function by solving this optimal problem with constraints. The term support vectors refers to the samples with non-zero Lagrange multiplier.

The support vectors are the input samples that will be used in the regression analysis, and they determine the SVR function. In other words, the non support vectors in the data set will not have any influenced in the regression model. Usually, the number of support vectors is small relative to the total number of samples.

Mathematically, the constrained quadratic programming problem of SVR can be expressed as (Wang et al., 2006):

$$\min_{w,b,\xi,\xi^*} \frac{1}{2} W^T W + C\sum_{i=1}^{n} \xi_i + C\sum_{i=1}^{n} \xi_i^*$$

subject to

$$W^T \phi(x_i) + b - z_i \le \varepsilon + \xi_i$$
$$z_i - W^T \phi(x_i) - b \le \varepsilon + \xi_i^*$$
$$\xi_i, \xi_i^* \ge 0, i = 1, \ldots, n$$

where W is the solution of the primal problem and C is the parameter of regulation that makes the trade off between margin ξ_i and prediction error ξ_i^*. Let $\{(x_1, z_1), \ldots, (x_n, z_n)\}$ denote a set of the training data, with the input $x_i \in R^n$ and the output $z_i \in R^l$. $\phi(x_i)$ is a non-linear function that maps the input feature into the higher dimensional space and ε is the error probability which limit the deviation between the regression function from the target. For a smaller value of ε, more support vectors are needed.

The dual problem for this quadratic programming problem is:

$$\min_{\alpha,\alpha^*} \frac{1}{2} (\alpha - \alpha^*)^T Q(\alpha - \alpha^*) + \varepsilon \sum_{i=1}^{l} \left(\alpha_i + \alpha_i^*\right) + \sum_{i=1}^{l} z_i \left(\alpha_i - \alpha_i^*\right)$$

Subject to

$$\sum_{i=1}^{l} \left(\alpha_i - \alpha_i^*\right) = 0, 0 \le \alpha_i, \alpha_i^* \le C, i = 1, \ldots, l$$

where $Q_{ij} = K(x_i, x_j) = \phi(x_i)^T \phi(x_j)$. The SVR regression function can be obtained as $\phi(x) = \sum_{i=1}^{l} \left(-\alpha_i + \alpha_i^*\right) K\left(x_i, x\right) + b$, where $K(x_i, x)$ is the kernel function. The required support vectors are the input vectors which have their corresponding α of non-zero value. It should be noted that this kernel function can assume different forms, such as linear function, polynomial function, radial basis kernel function, etc. The performance of the SVR depends on the setting of the kernel functions as well as their corresponding parameters.

2.2.10 Hybrid Intelligent Techniques

Hybrid intelligent technique is the approach that employs a hybrid combination of methods from artificial intelligence subfields. It includes for example hybrid systems of neural networks – evolutionary algorithms – fuzzy systems, hybrid multi-agent systems, knowledge-based neural networks, hybrid optimization techniques, hybrid machine learning models using SVM, Bayesian networks, statistical learning etc. Hybrid intelligent systems have been proposed for the data mining of different data sets (see for example Ao, 2003a, b, c, 2006, 2007; Negoita et al., 2005; Ao et al., 2008; Castillo et al., 2008). Hybrid intelligent systems are also very helpful for the genomics analysis. Yang (2008) proposed an intelligent decision system using machine learning techniques and markers to characterize tissue as cancerous, non-cancerous or borderline. Bosl (2007) presented the hybrid intelligent systems needed for modeling complicated biochemical dynamics using rule-based models to represent expert knowledge in the context of cell cycle regulation and tumor growth. Huang, Lee and Ho (2007) proposed an efficient evolutionary approach to gene selection from microarray data which can be combined with the optimal design of various multiclass classifiers. The proposed method has three hybrid components which are fully cooperated: an efficient encoding scheme of candidate solutions, a generalized fitness function, and an intelligent genetic algorithm (IGA).

2.2.11 Machine Learning Software

The Weka software package is a collection of machine learning algorithms in Java. The algorithms can be applied directly, or called from user's own Java code. The Weka package contains tool for data-processing, classification, regression, clustering, association rules and visualization. It is also suitable for developing new machine learning schemes. Weka is open source software available at:

http://www.cs.waikato.ac.nz/~ml/weka/

The Sleipnir C + + library (Huttenhower et al., 2008) implemented a variety of machine learning and data manipulation algorithms with the focus on heterogeneous data integration and efficiency for very large biological data collections. Sleipnir allows microarray processing, functional ontology mining, clustering, Bayesian learning and inference, and Support Vector Machine tasks to be performed for heterogeneous data. Source code (C + +) and documentation are available at http://function.princeton.edu/sleipnir

Matlab is a popular commercial high-level technical computing language and interactive environment for algorithm development, data visualization, data analysis, and numeric computation. Matlab is developed by MathWorks Company, and more information about Matlab is available at its company web page:

http://www.mathworks.com/

The SVM Light V6.01 is an implementation of Support Vector Machines in C, and the software also provides methods for assessing the generalization performance efficiently. It is available free-of-charge at:

http://www.cs.cornell.edu/people/tj/svm_light/

The Netlab library includes software implementations of a wide range of data analysis techniques, many of which are not yet available in standard neural network simulation packages. The Netlab library is available at:

http://www.ncrg.aston.ac.uk/netlab/index.php

2.3 Clustering Algorithms

The goal of the clustering algorithms is to figure out the underlying similarities among a set of feature vectors $\boldsymbol{x}$, and to cluster similar vectors together (Theodoridis and Koutroumbas, 2003). The clustering process can also be called unsupervised pattern recognition. This is different from the supervised pattern recognition, in which a set of training data is available, and which the classification algorithms can exploit this known information in advance. The clustering algorithms have many different applications in social sciences, engineering and medical science. In our case study, the algorithms are designed for clustering similar SNPs and selecting tag SNPs among these clusters.

2.3.1 Reasons for Employing Clustering Algorithms

The clustering process can be viewed as a combinatorial problem of putting the data points into optimal clusters. However, it is NP-hard to enumerate all such possibilities of clustering. Let $S(N, m)$ be the number of all possible clustering of N vectors into m groups. We can easily see that $S(N, 1) = 1$, $S(N, N) = 1$, and $S(N, m) = 0$, for $m > N$. It satisfies the following recursive relationship (Spath, 1980):

$$S(N,m) = mS(N-1,m) + S(N-1,m-1)$$

Its solution is found to be the Stirling numbers of the second kind (for details, see (Liu, 1968) etc.):

$$S(N,m) = \frac{1}{m!}\sum_{i=0}^{m}(-1)^{m-i}\binom{m}{i}i^N$$

We can see that the solutions for this problem explore exponentially with (Spath, 1980): $S(15,3) = 2.38 \times 10^6$, $S(20,4) = 4.52 \times 10^{10}$, and $S(100,5)$ is of order 10^{68}. It is impractical to enumerate all possible clusters for all possible values of m.

The clustering algorithms can also allow us to infer some hypothesis concerning the nature of the data. It can be a tool for suggesting hypothesis (Theodoridis and Koutroumbas, 2003). These hypotheses can be verified by using other data sets as validation sets. Another use is on the prediction that bases on groups. The algorithms can provide us with clusters that are characterized by the similarity of vectors within each cluster. When a new data set or pattern is available, we can assign it to the known cluster by comparing its characters with each cluster's characters. Clustering is important for the data reduction purpose too. There are many times that the amount of the data is very large and it is expensive to process all the data. Cluster analysis can be employed for grouping the data into a number of clusters, and then we can process each cluster as a single element.

2.3.2 Considerations with the Clustering Algorithms

In view of the computational difficulty, different clustering algorithms have been developed so that only a small number of the different possible combinations of the clusters will be considered. There are four main types of clustering algorithms: sequential algorithms, hierarchical clustering algorithms, clustering algorithms with cost function optimization, and others (like branch and bound algorithms, and genetic clustering algorithms). Different clustering algorithms usually produce different clustering results. It may depend on the problem to decide which type of clustering algorithms is employed.

In designing the clustering algorithm for solving a problem, a major issue is on how to define the similarity between two feature vectors. It is important to choose an appropriate measure for this task. Then, it is also important to choose an appropriate algorithmic scheme that clusters the vectors, basing on the selected similarity measure. Generally speaking, different results can be obtained with different algorithmic schemes. Expert opinions are often needed for the interpretation of the results and for choosing a suitable scheme. In our project, experts from the Genome Research Centre have provided us expert opinions on this subject.

2.3.3 Distance Measure

The concept of distance measure is important for the clustering process, which need this measurement of the mathematical distance between individual observations, and groups of observations (Finch, 2005). Distance in this context can be in the Euclidean sense, or some other comparable conceptualization like Manhattan distance, Hamming distance etc. This will affect the shape of the clusters, as some objects may be close to one another with one distance while further away with another distance. A primary assumption underlying these distance measures is that the variables are continuous in nature. Finch discussed about the distance measures in Cluster analysis with dichotomous data. The definition of a distance measure has an important role in the evaluation of clustering algorithms of gene expression profiles. Ido et al. (2007) compared different clustering solutions when using the Mutual Information (MI) measure, Euclidean distance and Pearson correlation coefficient. And, details about the quantitative comparison of how close the vectors are in the clustering process are available in the section of similarity measures in Chapter 4.

2.3.4 Types of Clustering

Clustering process can be grouped as hierarchical or partition clustering. Hierarchical clustering find successive clusters based on previously established clusters. Partition clustering establishes all clusters at once. It has been applied for exploring gene expression data in study like Heyer et al. (1999).

2.3.4.1 Hierarchical Clustering

Hierarchical clustering can be further divided into two basic groups: agglomerative and divisive clustering. Agglomerative clustering is of the bottom-up approach, that is, starts with each object as a separate cluster and then merge the objects into successively larger clusters. On the other hand, the divisive clustering is of the top-down approach. It starts with the whole set and then divide it into successively smaller clusters. Co-clustering is a clustering technique that not only clusters the objects themselves, but also their features as well. Another difference among the clustering algorithms is to look at whether the clustering process uses symmetric or asymmetric distances. For example, Euclidean distances are symmetric, which mean distance from object H to K is the same as the distance from K to H. In applications like sequence-alignment methods, asymmetric distances have been used (for example, Prinzie and Van den Poel, 2006).

2.3.4.2 Partition Clustering

The partition clustering algorithms usually determine all the clusters at once. K-means clustering, fuzzy c-means clustering and derivatives are popular partition algorithms. In the K-means algorithm, initially there are *k* randomly generated clusters. The average of all the points in a cluster is assigned as the center. Then, the data point is assigned to the nearest cluster center, and the new cluster centers are computed. The above two steps are repeat until the convergence criterion is met. In fuzzy clustering, each data point belongs to a cluster to a certain extend, rather than to one certain cluster completely. Besides this difference, the fuzzy c-means algorithm is similar with the k-means algorithm. Like hierarchical clustering, the partition clustering algorithms are also very popular for the genomic analysis, with microarray data sets etc.

2.3.4.3 Spectral Clustering

In spectral clustering, the dimensionality reduction for clustering in lower dimensions is performed with the spectrum of the similarity matrix of the data. A popular spectral clustering is the Shi-Malik algorithm, which is widely used for image segmentation. Liu et al. (2008) applied the spectral clustering to the analysis of correlation mutations in HIV-1 protease. The spectral clustering of the resulting covariance matrices disclosed two distinctive clusters of correlated residues. Oliveira and Seok (2008) proposed a multilevel spectral algorithm which can identify protein complexes more accurately with less computational time.

2.3.5 Clustering Software

Many commercial software packages such as Matlab have their own clustering functions. Nevertheless, the clustering algorithms are also available in various open source projects. For example, the software Cluster is an open source clustering software available with the implementation of the most commonly used clustering methods for gene expression data analysis. There are routines in the C clustering library, which enable the users to link with other C program. The software is available at:

http://bonsai.ims.u-tokyo.ac.jp/~mdehoon/software/cluster/software.htm

The following web page (http://astro.u-strasbg.fr/~fmurtagh/mda-sw/) contains the list of software and resources for clustering analysis by Fionn Murtagh.

CLUSTAG and WCLUSTAG are the clustering algorithms for finding tagging SNPs, with free downloads available at:

http://hkumath.hku.hk/web/link/CLUSTAG/CLUSTAG.html

http://bioinfo.hku.hk/wclustag/

2.4 Graph Algorithms

2.4.1 *Graph Abstract Data Type*

The graph algorithms have been applied in different domains like enumeration, geometry, topology, statistics, logistics, and computing etc. In graph theory, a graph G refers to the collection of a finite non-empty set $V(G)$ of elements called vertices (or nodes) and a finite set $E(G)$ of distinct pairs of distinct elements of $V(G)$ called edges. These two sets $V(G)$ and $E(G)$ are called the vertex set of and the edge set of G respectively (Beineke and Wilson, 1997). If the edge set consists of ordered pairs of distinct vertices, the graph is called directed graph. Similarly, if the edge set consists of unordered pairs of distinct vertices, it is undirected graph (Ahuja et al., 1993).

In a graph G, an edge e that connects two vertices v and w can be represented as $\{v, w\}$, or in short, vw. When there is the edge e that connects vertices v and w, these two vertices are called adjacent, and w is a neighbour of v. The neighbourhood $N(v)$ of a v is defined as the set of all vertices of G adjacent to v. A sequence of edges of the form $v_0v_1, v_1v_2, \ldots, v_{k-1}v_k$, where the vertices and the edges are all distinct, is called a path. A graph G is connected if there exists a path joining each pair of vertices of G. A disconnected graph is a graph that is not connected. A graph $G' = (V', E')$ is a subgraph of $G = (V, E)$ if $V' \subseteq V$ and $E' \subseteq E$. We can divide the disconnected graph into maximal connected subgraphs and these subgraphs are called components.

An edge $\{u, v\}$ in a graph G is said to cover its incident vertices u and v. For a graph $G = (V, E)$ with $E' \subseteq E$, E' is said to be an edge cover of G (or, to cover G) if, for each vertex $v \in V$, there exist at least one edge in E' which covers v (Foulds, 1991). The cardinality of the edge cover with the least number of elements is called the edge covering number of the graph G. A covering is said to be minimal if none of its proper subsets is a covering.

A vertex in a graph G is said to dominate those other vertices in G with which it is adjacent. For a graph $G = (V, E)$ with $E_1 \subseteq E$ and $U \subset V$, $E_1(U)$ is said to be an vertex dominating set for G if every vertex of V either belongs to $E_1(U)$ or is dominated by an vertex of U. The cardinality of the vertex dominating set with the least number of elements is called the vertex dominating number of G. A dominating set of a graph G is said to be minimum if there does not exist any dominating set of G with a smaller number of elements.

Among the graph techniques, the search algorithms are essential ones for attempting to find all the vertices of the graph that satisfy desired properties. Usually, a search algorithm will spread out from the source vertex and identify an increasing number of vertices that are reachable from the source. There are different rules for the search algorithms, like the breadth-first search, depth-first search, and reverse search algorithm etc.

2.4.2 Computer Representations of Graphs

Adjacency matrices and adjacency lists are two common computer representations of graphs. In the two-dimensional adjacency matrix, the rows and columns represent source and destination vertices. The entries in the matrix indicate whether an edge exists between vertices. The matrix can be a Boolean matrix such that the entry (i, j) is true if and only if edge (i, j) is in $E(G)$. For a graph G of n vertices and m edges, the adjacency matrix representation requires $O(n^2)$ storage, while the representation has constant-time lookup to check if an edge is in the graph. In an adjacency list, each node is represented as a data structure that contains a list of all adjacent nodes. For a graph G of n vertices and m edges, the adjacency matrix representation requires $O(m)$ storage, while the representation may require up to $O(n)$ time to check if an edge is in the graph. So, adjacency lists are preferred for sparse graphs. Otherwise, an adjacency matrix is a popular choice.

2.4.3 Breadth-First Search Algorithms

Breadth-first search (BFS) is an exhaustive search method that expands and checks all nodes of a graph systematically (Knuth, 1997). It begins at the root node and explores all the neighboring nodes. Then for each of the above nearest nodes, its unexplored neighbor nodes will be explored. This process will keep on until the goal is reached. For the time complexity, all the vertices and edges will need to be examined in the worst case, so the complexity is $O(|E| + |V|)$. BFS has been applied to find all connected components in a graph, and the shortest path between two nodes etc.

2.4.4 Depth-First Search Algorithms

Depth-first search (DFS) is a search method that starts at the root and explores as far as possible along each branch before backtracking (Knuth, 1997). In the graph case here, it may need to select some nodes as the root in the search, and then expand the first child node of this search tree, until it reaches a node of no children. Then, the search will return to the most recent node not yet explored. Its time complexity is the same as the BFS. DFS can be found in applications like finding connected components, topological sorting and solving puzzles with only one solution.

2.4.5 *Graph Connectivity Algorithms*

A cut of a connected graph G is a set of vertices whose removal makes the graph disconnected (Godsil and Royle, 2001). The size of a smallest cut is call the connectivity K(G). A graph is called k-connected when its connectivity is k or greater. Search algorithms like BFS can be applied to determine whether two vertices in a graph are connected or not. Undirected graph connectivity may be solved in O(log n) space.

2.4.6 *Graph Algorithm Software*

Public-domain sources for the program and data of The Stanford GraphBase are available at:

http://www-cs-staff.stanford.edu/~knuth/sgb.html

The Java source code of the Dijkstra's algorithm, a popular shortest path search algorithm, is available freely at:

http://renaud.waldura.com/doc/java/dijkstra/

The Stony Brook Algorithm Repository is a comprehensive collection of algorithm implementations for combinatorial problems. It is available for downloads by different programming languages and by different problem at:

http://www.cs.sunysb.edu/~algorith/

ABACUS is a software system in C + + for the branch-and-bound algorithms, available at:

http://www.informatik.uni-koeln.de/abacus/

2.5 Numerical Optimization Algorithms

Sometimes, optimal solutions are preferred after the transformation and reduction of the originally complex problems. The different optimization algorithms are very helpful for this purpose. Generally speaking, the optimization is to find out the solutions and their respective characterization for the following mathematical problems (Pardalos and Resende, 2002):

min $f(x)$

such that

$$x \in X \subseteq \Re^n$$

where the decision variables in the model is represented by x, the goodness of the solution is measured by the objective function f(x), and the set of feasible solutions are denoted by X. In solving the optimization problem, we are also concerned with the properties of the algorithms for solving the problem, for example, how fast the algorithm converges.

2.5.1 Steepest Descent Method

The steepest descent method is also called the gradient descent method. It starts at a point P_0, and moves from P_i to P_{i+1} by minimizing along the line extending from P_i in the direction of the local downhill gradient $-\nabla f(P_i)$, and so on until the termination goal is reached (Arfken, 1985). For functions of long, narrow valley structures, this method requires a lot of iteration for obtaining the solution.

2.5.2 Conjugate Gradient Method

Instead of using the local gradient for going downhill, the conjugate gradient method uses conjugate directions. Two non-zero vectors *u* and *v* are conjugate (with respect to a *n*-by-*n* symmetric, positive definite and real matrix A) if $u^T Av = 0$. The method was originally proposed by Hestenes and Stiefel (1952). For the case when the region near the minimum has the shape of a long, narrow valley, the method can finish the search much faster than the steepest descent method.

2.5.3 Newton's Method

Newton's method, also called Newton-Raphson method or the Newton-Fourier method, is a very popular root-finding algorithm. It can often converge quickly, especially if the iteration starts near the desired root. With an initial choice of the root's position x_0 for the function f(x), the algorithm is applied iteratively to obtain $x_{n+1} = x_n - f(x_n)/f'(x_n)$, where n = 1, 2, 3, … and f' denotes the derivative of the function f (Suli and Mayers, 2003).

2.5.4 Genetic Algorithm

Genetic algorithm (GA) is inspired by evolutionary biology with techniques like inheritance, mutation, selection, and crossover (recombination). It is a global search heuristics algorithm, and has applications in problems of bioinformatics and phylogenetics etc.

It requires a genetic representation of the solution domain and a fitness function to evaluate the solution domain. A pseudo-code GA algorithm is as followed (Vose, 1999):

1. Choose the initial population.
2. Evaluate the fitness of each individual in the population.
3. Iterate.
   3.1 Choose the highest-ranking individuals for reproduction.
   3.2 Generate the new offspring through crossover and mutation.
   3.3 Evaluate the fitness of each individual in the offspring.
   3.4 Replace poorest part of the population with the offspring.
4. Until the termination condition is reached.

2.5.5 Sequential Unconstrained Minimization

Sequential unconstrained minimization algorithm (SUMMA) is an iterative procedure for constrained minimization. The SUMMA refers to a general class of iterative algorithms that include, as particular cases, the barrier- and penalty-function method, the gradient descent method and the Newton method etc. At the kth step, we minimize the function $G_k(x) = f(x) + g_k(x)$ to obtain x_k, where the auxiliary functions $g_k(x) : D \subseteq R^J \rightarrow R_+$ are nonnegative on the set D, and each x_k is assumed to lie within D. The objective is to minimize the continuous function $f : R^J \rightarrow R$ over x in the set $C = \overline{D}$, the closure of D (Byrne, 2008).

2.5.6 Reduced Gradient Methods

The reduced gradient methods are implicit variable elimination algorithms for solving nonlinear programming (NLP) problems. An unconstrained function is first formed with a reduced number of $(N - K)$ variables, where the NLP problem has N design variable and K (where $K < N$) equality constraints. The constrained function is optimized for a solution of the design variables. As the other K variables are dependent on the chosen $(N - K)$ variables, their corresponding optimal values can be obtained using the equality constraints. The dependent variables which are eliminated from the optimization process are called the basic variables, while the remaining variables are called nonbasic variables (Kalyanmoy, 2004).

2.5.7 Sequential Quadratic Programming

The sequential quadratic programming (SQP) is the approach chosen by a number of software packages like MATLAB, OPTIMA and NPSOL etc. for the optimization problem. SQP can be regarded as a generalization of Newton's method for unconstrained optimization. SQP estimates the step away from the current point by minimizing a quadratic model of the problem. It replaces the objective function with the quadratic approximation and replaces the constraint functions by linear approximation. Boggs and Tolle (1995) provide a general survey of the SQP.

2.5.8 Interior-Point Methods

Interior-point methods have been inspired by Karmarkar's algorithm (Karmarkar, 1984) for linear programming. Different to some other popular methods, it achieves an optimal solution by traversing the interior of the feasible region, rather than around its surface. The predictor-corrector technique proposed by Mehrotra (1992) is the base of the current efficient implementation. The performance of the current implementations depends heavily on the efficient code for factoring sparse symmetric matrices.

2.5.9 Optimization Software

Generally speaking, the optimization tools are already included in a number of popular numerical software like Matlab and Mathematica. The Optimization Toolbox of the Matlab provides widely used algorithms for standard and large-scale optimization of both constrained and unconstrained problems. The toolbox includes functions for linear programming, quadratic programming, nonlinear optimization, nonlinear least squares, solving systems of nonlinear equations, multi-objective optimization, and binary integer programming. Mathematica's optimization tools are also very useful in optimization studies.

The NEOS Guide: Optimization Software provides a comprehensive listing of the optimization software with brief description:

http://www-fp.mcs.anl.gov/OTC/Guide/SoftwareGuide/

Arnold Neumaier has listed a number of global optimization codes in the public domain at his web site Global Optimization Software:

http://www.mat.univie.ac.at/~neum/glopt/software_g.html#bb_codes

Chapter 3
Advances in Genomic Experiment Techniques

3.1 Single Nucleotide Polymorphisms (SNPs)

3.1.1 Laboratory Experiments for SNP Discovery and Genotyping

The experimental designs in the laboratory for the SNP analysis can be grouped into two main types of methods (Kwok, 2002). The first ones are those for the detection of the mutation and SNP. These include methods like the denaturing high-performance liquid (dHPLC) (Xiao and Oefner, 2001), single-strand conformation polymorphism (SSCP) (Orita et al., 1989), conformation-sensitive gel electrophoresis (CSGE) (Gonen et al., 1999; Oto et al., 1993), and chemical cleavage (Ellis et al., 1998) and lastly the direct DNA sequencing. This direct DNA is in fact the gold standard for mutation detection and single nucleotide polymorphism discovery, even though it is relatively laborious in comparison with other methods (Taillon-Miller et al., 1999).

Another type of the methods is for the genotyping of the SNPs detected. These include: Taqman assay (Holland et al., 1991), single-base extension approaches (Hacia, 1999), pyrosequencing (Ahmadian et al., 2000; Alderborn et al., 2000), ligation (Nilsson et al., 2001), Invader assay (Ryan et al., 1999), primer extension with mass spectrometry detection (Karas and Hillenkamp, 1988; Buetow et al., 2001), and molecular beacons etc. The molecular beacons (Tyagi et al., 1998; Marras et al., 1999) work with the fluorescent colors, which are generated in sealed amplification tubes, for typing the single nucleotide polymorphism. The molecular beacons are especially suited for the SNP analysis, because their ability for the recognitions of the targets is of much higher specificity than traditional oligonucleotide probes. Besides the above discussion of the laboratory works for the discovery and the genotyping of SNPs, we will also talk briefly about the computational efforts for the discovery of SNPs in the following section.

3.1.2 Computational Discovery of SNPs

Various computational tools have been developed for the SNP discovery. This approach can be divided into four major steps (Barnes, 2003). First, in the identification step, highly similar sequences from individuals are figured out. This can be achieved with the BLAST program etc. The second step is to avoid the existence of spuriousness of the similarity by the program REPEATMASKER. The program can mask the high copy number repetitive elements. In the third step, the base-wise multiple alignments of the sequences are constructed (Marth et al., 1999). Lastly, the sequences from step 3 that are of base-to-base multiple alignments are scanned for nucleotide differences. Both freeware and commercial programs are available for this computational detection of SNPs. Among these, PolyBayes, PolyPhred and Sequencher are the popular ones.

3.1.3 Candidate SNPs Identification

A candidate SNP is a SNP that has a potential for functional effect. It includes SNPs in regulatory regions or functional regions, and even in some non-synonymous regions. There exist different methods for selection of such SNPs. It can be identified by eye with the human genome browsers. This can give us the detailed study but is limited to small focus loci only. On the other hand, automated programs can be employed too.

3.1.3.1 Gene-Finding Methods

In many experiments and analysis, it is necessary to have the information about the genes of a genome, which can be found with both laboratory methods and computational methods (Salzberg et al., 1998). Computational methods are comparatively fast and cheap, but may not be as accurate as the laboratory methods. In the laboratory sequencing, the complementary DNA (cDNA) sample is sequenced to generate the expressed sequence tag (EST). The cDNA comes from the RNA that is from the transcription of the genes, and that is extracted from the cytoplasm and copied back into DNA using an enzyme called reverse transcriptase. In this process, the RNA is captured only after the introns have been spliced out and thus it can give us a very accurate picture of the encoding regions. EST sequencing projects have already generated many ESTs from humans and these datasets provide direct biological evidence for the gene regions. There have been efforts for constructing assemblies of these gene fragments information to obtain a complete coding region sequence (Adams et al., 1995; Boguski and Schuler, 1995).

Among the computational methods, there are different approaches for gene finding. The direct approach is to locate exactly where the following four signals can be found. These signals are the start codon, the donor sites (the beginning of each intron), the acceptor sites (the end of each intron), and the stop codon. If all of these signals can be identified correctly, we can have the protein product accurately. Nevertheless, the identification of these signals itself can not be completely accurate.

There exist other methods for assisting the determination of the coding regions. Content scoring methods can analyze larger sequence regions for other statistical patterns. Many of these methods have utilized the observation that codon frequency is different in coding and non-coding regions. Another approach is to identify different regions with the computation of their respective entropy (information content). The content has been shown to be different among different types of regions. Data mining tools like neural network, hidden Markov model, decision tree and rule-based algorithms have been employed for this gene finding task.

3.1.3.2 Polymorphisms in Coding Regions vs. Non-coding Regions

In the candidate SNP identification process, it may be helpful to take into account whether the SNPs are in the coding regions or in the non-coding regions. This information about the coding regions (genes) and non-coding regions can be obtained with the above gene-finding methods. The non-synonymous changes in coding regions of genes cause the alternations in amino acid sequences. The amino acid variants have accounted for a large amount of diseases. These coding polymorphisms can affect the protein folding, active sites, protein-protein interactions and its stability. Thus, it is clear that the SNPs in the coding region are important for the analysis. Nevertheless, polymorphisms in other regions may be of significance too. For example, variants in regulatory regions may change the transcription factor binding sites. Polymorphisms in the untranslated regions (UTR) of mRNA may change its stability and, it may even be true that polymorphisms in the introns can change the splicing efficiency.

3.1.4 Disease Studies with SNPs

3.1.4.1 Main Disease Types

A common type of diseases is the Mendelian disorder, which is also called single-gene or monogenic disorder (Pevsner, 2003). Their main causes are mutations in single genes in the human genome. Examples of this kind of disorders include hemophilia A, color blindness and breast cancer (Scheuner et al., 2004).

Another common type of diseases is the complex disorder, such as the diabetes, high blood pressure, obesity and cardiovascular disease. They are caused by the defects that occur in multiple genes. The complex disease is also called multifactorial, meaning that they are caused by both genetic and environmental factors. Another characteristic of the complex diseases is that the susceptibility alleles are of high population frequency.

3.1.4.2 Relationships Between Diseases and SNPs

It is estimated that about 99.9% of two selected genomes of the same gender are the same. It is the remaining 0.1% differences in the genome that arise the DNA sequence variations. They determine the individuality among the human. Among these genetic variations, SNP is the most common genetic variation in the human genome. It is a challenging task to figure out the relationship between the SNPs and the various diseases (Wang and Moult, 2001). SNPs exist in both the coding region of the genes and other non-coding regions, and different kinds of the SNP variations can provide different useful information about the diseases in different ways:

1. Functional variation refers to the situation when the SNP is with a nonsynonymous substitution in a coding region;
2. Regulatory variation happens when the SNP is in a non-coding region, but it can influence the properties of gene expressions (Cowles et al., 2002);
3. Associations of the SNP with the disease become useful when there are some SNPs close enough to the mutations that cause the diseases. These SNPs can then be utilized in the association studies with the diseases (Sherry et al. 2000);
4. Construction of the haplotype maps becomes possible with the collection of the information of the SNPs. The map is helpful for selecting SNPs that can be informative for explaining the differences in different ethnic groups and populations.

3.2 HapMap Project for Genomic Studies

3.2.1 HapMap Project Background

Mutation databases of the human genome are becoming more and more important in different areas like health care (Taylor et al., 2005). The mutation databases can be divided into general mutation databases and locus-specific mutation databases (LSDBs). The general mutation databases include Online Mendelian Inheritance in Man (OMIM), Human Gene Mutation Database (HGMD), Human Genome Variation Database (HGVbase), dbSNP database and HapMap project database etc. For the LSDBs, the focus is on the variation within a single gene, and, it is usually run by medical experts in that particular gene or phenotype. It has been estimated that there exists over 270 LSDBs on the World Wide Web.

In the HapMap project, it has been planned to find out the genetic similarities and differences in human genomes (HapMap, 2005). It is also to compare the genetic sequences among different individuals for locating chromosomal regions where genetic variants are shared. In order words, the HapMap can be regarded as a catalog of common human genomic variants. With the availability of this information freely, it will enable the researchers to figure out genes involved in diseases and to estimate individual responses to medications and environmental factors.

The HapMap project is a multi-country collaboration of the scientists from Canada, China, Japan, Nigeria, the United Kingdom and the United States. The University of Tokyo and Health Sciences University of Hokkaido in Japan are responsible for 24.3% of the genome. Wellcome Trust and University of Oxford in United Kingdom are responsible for 23.7% of the genome. McGill University in Canada takes care of the 10.1% genome. The Chinese HapMap Consortium in China is responsible for 9.5%. Within this consortium, Hong Kong HapMap group is doing the genotyping of a total of 2.5% genome. Harvard, Johns Hopkins, MIT and UCSF etc., in United States are doing the genotyping of a total of 32.4% genome. Cold Spring Harbor Laboratory in New York does the role of the data coordination center, with the funding from the US National Institute of Health.

Different DNA samples were taken from blood samples of volunteer donors in the HapMap project. The donors of total 270 individuals come from the following populations: Han Chinese in Beijing (HCB), Japanese in Tokyo (JPT), Yoruba in Ibadan of Nigeria (YRI), and Utah residents of US with ancestry from northern and western Europe (CEU). The samples are identified by which population they come from, even though no medical or personal information was disclosed. Samples of different populations are needed here, even though it is true that, in one human population, we can find most of the common haplotypes in human chromosomes. It is because, the frequencies of any given haplotype may be different among different populations, and some new haplotypes may exist in just a single population. Thus, it becomes necessary to identify the haplotype information in various populations.

3.2.2 Recent Advances on HapMap Project

The project formally began in October 2002 (HNGRI, 2005), and it was planned at that time that, by September 2005, it would produce the map of common patterns of human genomic variation. By the end of February 2005, 7 month ahead of the target date, the group completed the first draft of the human haplotype map (HapMap). It consists of 1 million markers (SNPs) of genetic variations. The total genotyped SNPs for population CEU totaled 1,073,663 on 1st March 2005. The figure for HCB is 1,044,686, that for JPT is 1,044,416 and that for YRI is 1,034,205. They are genotyped with a SNP density of 1 every 5 kb in all populations under study. The data can be downloaded freely from HapMap's official site (www.hapmap.org). This can enable the scientists to begin analysis of the variations among individual genomes. This is impossible with just the human DNA genomic sequence, but come true when the consortium has compared different genomic sequences of different human beings. The consortium published the comprehensive analysis results data in 2005 (HapMap, 2005).

After this completion of the first draft in its first phase, the consortium's goal is set to get an improved version of the haplotype map of much higher density. This is achievable because of the development of the high-speed, high-throughput genotyping capacity technology. Perlegen Sciences, Inc., of Mountain View, California US, is responsible for the testing of the 4.6 million additional SNPs and this new information will then be merged with the current map. This effort becomes possible with the grant of US$6.1 million from the National Human Genome Research Institute (NHGRI). NHGRI is part of the National Institutes of Health (NIH). In this second phase, the density of genotyping the SNP increases fivefold in the human genome. By its completion, it is estimated that every known catalog of human variation on the HapMap samples can virtually be tested.

In July 20, 2006, the HapMap project released its phase II dataset, which contains genotypes, frequencies and assays for bulk download. The data also includes genotypes from the Affymetrix 500k genotyping array. In the phase II, there existed 3.28 million non-redundant SNPs for CEU population, 3.31 million for the CHB + JPT population, and 3.24 million for the YRI population (HapMap, 2006). The preliminary release of HapMap Phase 3, containing genotype and pedigree information for 11 populations (including individuals in the original four from earlier phases of the project), is available on May 27, 2008.

3.2.3 Genomic Studies Related with HapMap Project

With the progress of the HapMap project, this can enable the researchers to have more information about the SNPs in the human genome. This leads to a number of publications related with the HapMap project. For example, the publication (International HapMap Consortium, 2005) is about the HapMap data, which contain the information of more than 1 million SNPs in the human genome, the recombination hotspots, a block-like structure of linkage disequilibrium and low haplotype diversity. It also showed how the HapMap can assist the design and analysis of genetic association studies. Thorisson et al. (2005) has presented guides for using the different tools available in the HapMap web page.

Clark et al. (2005) studied the ascertainment bias with the HapMap dataset, and concluded that its effect on the power erosion of association tests will likely be small. Bakker et al. (2005) has investigated the Tag-SNP selection for genome-wide association studies. It found that the power is robust to the completeness of the reference panel where the tags are selected. Myers et al. (2005) conducted the statistical analyses of genetic variation data for a high-resolution genetic map of the human genome. More than 25,000 recombination hotspots were found in the study. Smith et al. (2005) applied the HapMap dataset for the study on the relationship between sequence features and the degree of linkage disequilibrium in the genome. It was noticed that the variation in LD is roughly similar across populations. Nevertheless, the study of Weir et al. (2005) found that there exists substantial heterogeneity of genetic population structure among the populations of HapMap, even though there was also similarity between them.

3.3 Haplotypes and Haplotype Blocks

3.3.1 Haplotypes

The specific set of alleles in a region of a single chromosome that has not been broken up by recombination is called haplotype. The haplotype regions are separated by regions of recombinations. The haplotypes in the human genome are the products of the reproduction and thus are determined by the history of the population. New haplotype can be formed with additional mutations or by recombination of the parental chromosomes. Because of the limitation of their existing time, these new haplotypes usually have not spread widely across different populations and are restricted to their original population.

It has been estimated that the number of generations since the most recent common ancestor of any two humans are of order 10^4 generations (Intl. HapMap Consortium, 2003). Comparatively, the mutation rate is very low, of the order 10^{-8} per site per generation, so nearly every variable site is the results of a single historical mutational event. As a result, each new allele is initially associated with the other alleles of the same chromosomal background. Then, segments of the chromosome arc shuffled through the recombination events generations after generations.

The following example is a simple illustration of the formations of the haplotypes. Let the two ancestral chromosomal regions concerned be represented by MMMMMMMMMM and FFFFFFFFFF. After many generations of recombination events, we study five of the child chromosomal regions here:

1. MMMFFMMMMF
2. MFFFMMFFFM
3. MMMFMMMMMF
4. MFFMMMMFFM
5. MMFFMMFFMM

Assume that there exists a disease D in chromosome 1 and 3 and the disease gene is located between the second and third position. Then, we can see that the MMM segment of the first to third position of the chromosome 1 and that of the chromosome 3 are identical and the correlation between the disease D and these MMM alleles are 100% in this example.

3.3.1.1 Haplotypes and Linkage Disequilibrium

The association studies are traditionally done with the individual genetic markers for computing the linkage disequilibrium (LD). Nevertheless, these traditional approaches often give results of an erratic and non-monotonic picture. SNPs have become the promising markers for association studies. Daly et al. (2001) began the studies of the haplotypes for the LD analysis and compared these results with the results from single-marker LD. The study is on samples of 129 trios from a European-derived population. It is shown that the noises, which are presumably caused by the marker history etc., disappear when using the haplotype-based LD. Daly's results also show that there exists a picture of discrete haplotype blocks that are of order tens to hundreds of kilobases. Inside each block, there is only a little diversity, while between the blocks there are punctuations that show the potential sites of recombination.

Daly et al. have observed that, over a long distance, most haplotypes can be cataloged into a few common haplotype categories. Thus, Daly et al. employed the Hidden Markov Model (HMM) for their study. Daly et al. defined the haplotype categories as the states of the HMM and assigned observed chromosomes to those hidden states. The transition probability in each interval is estimated with an EM algorithm. With the method, the maximum-likelihood assignment to the haplotype categories for each position can be obtained. The maximum-likelihood estimates of the historical recombination can also be found.

The LD between pairs of markers can be calculated with the standard measures like *D*' (Lewontin, 1964) and r^2 (Hill and Robertson, 1968; Ohta and Kimura, 1969). Methods like 'sliding window' LD profiles (Dawson, 2000), LD unit maps (Maniatis, 2002), haplotype blocks and estimations of meiotic chromosomal recombination rates (Hudson, 1987; Fearnhead and Donnelly, 2001) are being developed for the identification of the high LD and haplotypes in chromosomes. After the analysis of the LD with the HapMap data in the first phase, low LD regions will be identified and further genotyping may be needed in these regions for the genetic details.

3.3.1.2 Measuring the Haplotype Diversity

Clayton (2001) studied the measuring of the haplotype diversity of the diallelic cases. Diallelic allele means that we only take into consideration if it is a major or minor allele, and we can represent them with codes 0 and 1. Define each observation, $i = 1, \ldots, N$, of a haplotype, as a vector $z_i = \{z_{ij}, j = 1,\ldots,S\}$ of alleles, where S is the total number of linked polymorphic markers. Clayton defined the locus diversity as the total number of differences between all N^2 pair-wise comparisons between the observations (Escoffier, 2001). For the locus j, the diversity is as followed:

$$D_j = \sum_{i=1}^{N}\sum_{k=1}^{N}\left(z_{ij} - z_{kj}\right)^2 = 2\left\{N\sum_{i=1}^{N} z_{ij}^2 - \left(\sum_{i=1}^{N} z_{ij}\right)^2\right\}$$

where the difference between two alleles is $(z_{ij} - z_{kj})$ and equals zero if the observations i and k are the same. It equals ±1 if they differ. Assume that the number of chosen htSNPs is H. Let G denote the total number of groups that are defined by haplotypes of these htSNPs. And denote these ht-haplotype groups as H_g, $g = 1, \ldots, G$. Then, the residual diversity for the locus j is:

$$R_j = \sum_{g=1}^{G}\left\{\sum_{i\in H_g}\sum_{k\in H_g}\left(z_{ij} - z_{kj}\right)^2\right\}$$

The proportion of diversity "explained" (PDE) by the set of htSNPs, at the locus j, is given by:

$$P_j = 1 - \frac{R_j}{D_j}$$

Similarly with the above definitions at the individual locus *j*, the diversity for the haplotype as a whole can be defined as:

$$D = \sum_{i=1}^{N}\sum_{k=1}^{N}(z_i - z_k)^T(z_i - z_k) = 2\left\{N\sum_{i=1}^{N} z_i^T z_i - \left(\sum_{i=1}^{N} z_i\right)^T\left(\sum_{i=1}^{N} z_i\right)\right\}$$

and the residual diversities for overall as:

$$R = \sum_{g=1}^{G}\left\{\sum_{i\in H_g}\sum_{k\in H_g}(z_i - z_k)^T(z_i - z_j)\right\}$$

The overall proportional of diversity explained is given by:

$$P = 1 - \frac{R}{D}$$

Clayton has implemented the above computation of the haplotype diversity in the program hapdiv. It is written in a statistical macro language called Stata. In its default setting, thc subset size is set to a maximum number of five. If a larger number of subsets are set, the computation time will become longer and may even be infeasible for exhaustive subset search.

3.3.2 *Haplotype Blocks*

The idea of the haplotype blocks has come from studies like that of Gabriel et al. (2002). Gabriel et al. showed that the human genome can be divided into haplotype blocks, which are defined as regions of little historical recombination and of only a few common haplotypes. Gabriel et al. defined a haplotype block as a region over which only a small percentage (<5%) of comparisons among informative SNP pairs show evidence of historical recombination. The 5% tolerance level is chosen as many other biological forces besides recombination can disrupt haplotype patterns. For example, these biological forces can be from recurrent mutation, gene conversion etc.

The study of Gabriel et al. (2002) is a part of the SNP Consortium Allele Frequency Projects. In their work, the haplotype patterns across 51 autosomal regions were characterized. Gabriel et al. used samples of the European, Asian and African American from the Coriell Cell Repository (http://locus.umdnj.edu/ccr/). The expectation-maximization (EM) algorithm by Excoffier and Slatkin was employed for the haplotype frequencies within the blocks. Gabriel et al. addressed the questions of the average size of the haplotype blocks and studied the genetic details in each block, like the size and diversity of haplotypes within the blocks.

Another method for studying the genetic details is the construction of the LD maps of the genome. The first LD maps for identifying the hot and cold spots of recombination in the genome were proposed by Maniatis and colleagues (Maniatis et al., 2002). The maps are based on the Malecot equation. Maniatis et al. replaced the time variable (the number of generations) there with the distance variable (distance between two SNPs). The derivation of this LD map is parametric and requires the estimation of three coefficient parameters.

3.3.3 Dynamic Programming Approach for Partitioning Haplotype Blocks

Zhang et al. (2002) have developed a dynamic programming approach for the partitioning of the haplotype blocks. It is to minimize the number of haplotype tagging SNPs needed for accounting most of the common haplotypes in each block. The algorithm has the advantage that any measure of the haplotype quality can be used in the algorithm. Compared with the greedy method used in study of Patil et al., the number of htSNPs identified by the dynamic programming is 21.5% smaller. The number of blocks is also smaller, with a reduction of 37.7%.

In the dynamic algorithm, Zhang et al. define the common haplotype as those that occur more than once in a block. It is required that a significant percent of the haplotypes in each block are common haplotypes. With the blocks defined, the target is to minimize the number of htSNPs that can distinguish at least α percent of the haplotypes in the block. α is called the coverage of the htSNPs. This minimization problem is known as Minimum Test Set problem, and it is shown to be a NP-complete problem (Garey and Johnson, 1979).

Mathematically, let $r_1, r_2, \ldots, r_n$ be the SNPs (Zhang et al., 2002b). A Boolean function block can be defined as block $(r_i, r_{i+1}, \ldots, r_j) = 1$ if at least α percent of the haplotypes formed by the SNPs $r_i, r_{i+1}, \ldots, r_j$ are presented more than once. Otherwise, the block $(r_i, r_{i+1}, \ldots, r_j)$ is set to a value of zero. Let $f(\cdot)$ be the number of htSNPs in a block. Then, when the sequence is partitioned into blocks $B_1, B_2, \ldots, B_I$, the total number of htSNPs are given by . Let S_i be the number of htSNPs for the optimal block partition of the first j SNPs, $r_1, r_2, \ldots, r_n$, and define $S_0 = 0$. Zhang et al. have developed the recursive algorithm of the dynamic programming as:

$$S_j = \min_{1 \le i \le j} \{ S_{i-1} + f(r_i, \ldots, r_j), block(r_i, \ldots, r_j) = 1 \}$$

When there exist several block partitions of equal minimum number of htSNPs, the one with the minimum number of blocks is preferred. Let C_j be the number of minimum number of blocks of all block partitions that require S_j htSNPs. Then, another recursive step can be applied:

$$C_j = \min_{1 \le i \le j} \left\{ C_{i-1} + 1 : block(r_i, \ldots, r_j) = 1, S_j = S_{i-1} + f(r_i, \ldots, r_j) \right\}$$

This recursive step can compute the minimum number of blocks in the partition.

3.4 Genomic Analysis with Microarray Experiments

In each eukaryotic cell, it is estimated that there are between 5,000–60,000 protein-coding genes (Pevsner 2003). At any time during the transcription process of DNA into RNA, only a subset of the genes is expressed as mRNA transcripts. These are also called gene expression and the set of expressed genes in the genome are sometimes called the transciptome. There are various ways for the regulation of the gene expression, like by the regions, the development stages, disease states and gene activity etc.

3.4.1 Microarray Experiments

Generally speaking, microarray is a solid substrate where the DNA is attached to in an ordered manner at high density (Geschwind and Gregg, 2002). Among the high-throughput methods of gene expression, the microarray has been the most widely used one for assessing the differences in mRNA abundance in the biological samples. The mRNAs are produced in the first step of the gene expression, the transcription process, and the mRNAs are relatively simple to study in high-throughput modes. With the work of Patrick Brown and his colleagues (DeRisi, 1996), microarray has been gaining its popularity.

In a single microarray experiment, the expression levels of as many as thousands of genes can be measured simultaneously. Thus, it can enable the genome-wide measurement of gene expression. This is a large improvement over the situation of "one gene per experiment" in the past. As a result, microarray has been found useful for different types of biological researches, for example, tissue-specific gene expression, developmental genetics, genetic diseases, complex diseases, and environmental monitoring etc.

A typical microarray experiment consists of the following five steps (Amaratunga and Cabrera, 2004):

(a) Preparation of the microarray: Drops of purified single-stranded DNAs is placed onto glass microscope slide.
(b) Preparation of the labeled sample: mRNAs are purified from the sample contents, and then reverse-transcribed into more stable cDNA or cRNA.
(c) Hybridizing of the labeled sample: Label sample is then sealed in hybridization chamber for hybridization reactions.
(d) Scanning of the microarray: This is to check the amount of labeled sample bound to each spot of the microarray.

(e) Data analysis of the scanned image: The scanning product of the microarray is a gray scale image, and image-processing tools are needed to convert the image into spot intensity measurements for further data analysis.

Multi-channels cDNA microarray and oligonucleotide array (pioneered by Affymetrix) are two popular microarrays. New microarray technologies are also emerging, for example, the bead-based microarray technology.

3.4.2 Advances of Genomic Analysis with Microarray

With the advances in DNA microarray technology (Causton et al., 2003), we can have the gene expression values at different time points of a cell cycle. In the simplest case of time series expression analysis, two time points are taken: before and after an event. A more comprehensive study will involve the taking of values at different periods. The frequencies of the time points can have ranges from several minutes to several hours.

Various methods like self-organizing maps (Nikkilä et al., 2002), *k*-nearest neighbor (Acta, 2001) and hidden Markov models (Ji et al., 2003) have been employed for the microarray analysis. These studies mainly focus on the clustering and the measurement of the similarity among the different expressions. For the gene expression time series analysis, methods like warping algorithms (Aach and Church, 2001), the comparison of similarity functions of the genes (Butte et al., 2001), the identification of gene regulatory networks with graph method (Chen et al., 2001), and dynamic models (Dewey, 2002) etc., have been developed.

In the literature, different methods have been developed to analyze gene expression time series data, see for instance (Costa et al., 2002; Yoshioka and Ishii, 2002; Tabus and Astola, 2003; Syeda-Mahmood, 2003; Wu et al., 2003; Jiang et al., 2003; Futschik and Kasabov, 2002; Kesseli et al., 2004; Tabus et al., 2004; Sakamoto and Iba, 2001; Zhang et al., 2003; Craig et al., 2002; Langmead et al., 2002, etc.). However, their approaches are different from our proposed PCA-NN modeling for gene expression time series. About half of these researches have their focus on building special clustering algorithms for these time series data, while the others have tackled the problems of inferring systems of linear differential equations, the visualizing of the gene data and the determination of their periodicity.

As said above, some special clustering algorithms have been employed to explore the gene expression time series data from the microarray experiments. Costa et al. (2002) have proposed the symbolical description of multiple gene expression time series. Each variable will take as a set of values in a time series and the results are compared with Self-Organizing Map algorithm. Yoshioka and Ishii (2002) have employed a clustering method based on mixture of constrained PCA. It can classify genes with similar expression patterns into the same cluster regardless of their magnitude (scale). In the study (Tabus and Astola, 2003), Tabus and Astola have handled the problem of the non-uniformly sampling of the gene expression time series. The minimum description length model is fitted to each gene and then the optimum parameters are used for clustering the genes. The extrapolation of the gene expression time series data by the minimum description length model can be applied in our methodology too for non-uniformly sampling data.

Syeda-Mahmood (2003) has studied a clustering algorithm that uses the scale-space distance as a similarity metric. The scale-space analysis is to detect the sharp twists and turns of the gene time series and to form the similarity measure between time profiles. Wu et al. (2003) have developed a procedure for the determination of the minimal number of samples or trials required in a microarray experiment for clustering. The procedure is an incremental process that will terminate when the evaluation of the results of two consecutive experiments of k-means clustering shows they are sufficiently close. Jiang et al. (2003) use a density-based approach to identify the clusters such that the clustering results are of high quality and robustness. Futschik and Kasabov (2002) employ the fuzzy c-mean (FCM) clustering to achieve a robust analysis of gene expression time series. The issues of parameter selection and cluster validity are also addressed.

3.4.3 Methods for Microarray Time Series Analysis

The construction of genetic network from gene expression time series is tackled in (Kesseli et al. 2004; Tabus et al., 2004; Sakamoto and Iba, 2001). Kesseli et al. have employed monotonic time transformations (MTT) for inferring a Boolean network. Several different methods of clustering have been used to form different transformations. Tabus et al. build systems of differential equations for specifying the genetic networks. The structure of the networks is inferred by operating with the exact solutions of the linear differential equations, which are obtained through the eigenvalue decomposition of the system matrix. Sakamoto and Iba also use a system of ordinary differential equations as a model of the network and infer their right-hand sides by using genetic programming (GP) instead. The least mean square (LMS) method is used along with the GP to explore the search space more effectively in the course of evolution. In these systems of linear differential equations, there is a strong assumption that the genetic interactions are linear. Instead, with our PCA-NN algorithm, we can have the advantage of the nonlinear flexibility of the neural network. Also, we have employed the AIC test to decide the optimal lag length used in our models, whereas, in the above models, only one lag length of each gene expression value change is included. The lag length refers to the number of lags (the number of previous values of the variable) used in the model.

The visualizing of the gene expression time series is discussed in studies (Zhang and Zhang, 2003; Craig et al., 2002). Zhang et al. have introduced the first Fourier harmonic projection (FFHP) to translate the multi-dimensional time series data into a two-dimensional scatter plot. The spatial relationship of the points reflects the structure of the original dataset and the relationships among clusters become two-dimensional. Craig et al. propose the display technique that operates over a continuous temporal subset of the time series, with direct manipulation of the parameters defining the subset. Its advantage is that the number of elements being displayed will not be reduced.

Langmead et al. (2002) formulate the task of estimating an expression profile's periodicity and phase as a simultaneous bicriterion optimization problem. The maximum entropy-based analysis technique is employed for extracting and charactering rhythmic expression profiles, and is found to work better than the Fourier-based spectral analysis for signals in the microarray experiments. Yeang and Jaakkola (2003) explain time correlations between gene expression profiles through factor-gene binding information to estimate latencies for transcription activation. This can estimate latencies for transcription activation. The resulting aligned expression profiles are subsequently clustered and again combined with binding information to determine groups or subgroups of co-regulated genes.

Chapter 4
Case Study I: Hierarchical Clustering and Graph Algorithms for Tag-SNP Selection

More than 6 million single nucleotide polymorphisms (SNPs) in the human genome have been genotyped by the HapMap project by the end of July 2006 (HapMap, 2006). Although only a proportion of these SNPs are functional, all can be considered as candidate markers for indirect association studies to detect disease-related genetic variants. The complete screening of a gene or a chromosomal region is nevertheless an expensive undertaking for association studies. A key strategy for improving the efficiency of association studies is to select a subset of informative SNPs, called tag SNPs, for analysis (Johnson et al., 2001).

4.1 Background

4.1.1 Motivations for Tag-SNP Selection

The studies of the large-scale genotyping of single nucleotide polymorphisms can date back before the launch of the HapMap Project. Wang et al. (1998) studied the feasibility of the large-scale genotyping of single-nucleotide polymorphisms in the human genome with a large-scale survey of 2.3 megabases of human genome. Among these 2.3 megabases, there exist about 3 Mb of protein coding regions and this is about 2.5% of all the coding regions of human genes, with total length of about 120 Mb in different parts of the genome. They did the examination with gel-based sequencing and high-density variation-detection DNA chips, and developed prototype genotyping chips that can simultaneously genotype 500 SNPs. In their study, a total of 3,241 candidate SNPs were identified and a genetic map was constructed that could show the location of 2,227 out of these 3,241 SNPs. Wang et al. demonstrated that the large-scale identification of human SNPs is feasible in this work.

With the progresses of different SNPs projects like the HapMap project, results of the large-scaled genotyping of SNPs in genome have been becoming available. As it is very expensive to genotype all these SNPs for the disease analysis each time, it becomes necessary to select the tag SNPs for these analysis. Nevertheless, the selection of the tag SNPs is not a simple task. Methods like random selection and evenly-space selection are not efficient, as empirical studies have shown that the extent of association between nearly markers can vary largely in different genomic regions (Jeffreys et al., 2001; Dawson et al., 2002; Taillon-Miller et al., 2000).

When SNP alleles co-inherit on the haplotypes, this can lead to the associations between these alleles in the population. This is what we call linkage disequilibrium (LD). As there is more likely recombination between two SNPs that are far away from each other, these associations between SNPs will usually decline with distance. There are many empirical studies that have shown strong associations (high LD) between nearby SNPs in the human genome (Jeffreys et al., 2001; Reich et al., 2001; Abecasis et al., 2001; Dawson et al., 2002). With the strong associations in a genome region, it means that only a few haplotypes or tag SNPs can explain most of the variation among people in the region. In other words, in some regions of strong LD, a small number of suitable SNPs are enough for representing the whole sets of SNPs in the regions, whilst, in region of low LD, more SNPs are required to serve as tag SNPs. With the suitable choice of the tag SNPs, it can minimize the cost of genotyping while at the same time retain as much information about the genetic variations as possible.

Recombination is a significant source for the breaking down of the linkage disequilibrium (Kwok, 2002). In regions with less recombination, the genetic variation there is usually much lower. In the CpG dinucleotide regions where the mutation is more likely to occur, the density of SNPs there is much higher than other regions of the genome. Only 1 to 2% of the genome is the CpG regions, but about 25 to 30% of the SNPs are found in these regions (Halushka et al., 1999). Thus, it is important to choose a suitable amount of tag SNPs for representing these regions of highly different SNP densities.

In the association studies of the diseases and the SNPs, the number for SNPs required can be in the range of several hundreds to thousands. Thus, it is necessary to develop cost-effective methods that can deal with the large-scale analysis of the diseases with SNPs.

4.1.2 Pioneering Laboratory Works for Selecting Tag SNPs

Johnson et al. (2001) have studied the effectiveness of choosing a small amount of SNPs for the haplotype tagging to identify common disease genes. Johnson et al. have 384 samples from European individuals and have scanned 135 kb of DNA from nine different genes. A total of 122 single nucleotide polymorphisms are genotyped. It is found that the knowledge of the common haplotypes and the haplotype tagging SNPs can explain the complex patterns of LD between adjacent markers. By working with the htSNPs, the genotyping cost can be reduced significantly (34 htSNPs are required for the total 122 SNPs in this work). Key fine-mapping data within regions of strong LD can also be obtained.

Johnson et al. have demonstrated that there is lack of correlation between the level of LD and the physical distance in regions of distances less than 100 kb.

If instead, the underlying haplotypes are characterized, it can be clearly defined for the relationships between the alleles in these regions. Johnson et al. have estimated that, for an initial gene-based cataloging of the htSNPs, all the regions of the genes and at least 3 kb of up- and downstream sequences of each gene should be sequenced in a minimum of 30 individuals. This would give a result of larger than 95% power for detecting all variants of frequencies of at least 5%.

In the Gabriel et al. study (2002), the statistical power of the haplotype framework is found to be substantial in the association studies of common genetic variation across each region. Gabriel et al. have found that the correlation of the SNPs inside each block is high and that only a fraction of the total SNPs in the block will be needed for testing in a medical research. A major attraction of the haplotype block methods is the ability to identify regions (blocks) where only a few common haplotypes can capture most of the genetic variation, such that only a small number of haplotype tagging SNPs (htSNPs) are required for genotyping. Gabriel et al. have estimated that for the human genome, a total of 300,000 to 1,000,000 well-chosen htSNPs would be needed. These pioneering works have leaded to efficient tag-SNP selection methods that are based on the determination of the haplotypes and haplotype blocks. It is found relatively easy to determine whether a haplotype block is associated with a disease or not. It is also found that a relatively small number of SNPs is enough for marking the common haplotypes in each block.

4.1.3 Methods for Selecting Tag SNPs

4.1.3.1 Types of Tag-SNP Selection Methods

As described in the previous discussion, there exist redundant information in the whole set of SNPs and it is expensive to genotype this whole set. Different approaches have been developed to reduce the set of SNPs that are to be genotyped. These selected subsets are called haplotype tagging SNP (htSNPs) or tag SNPs. The approaches can be divided into two main categories (CIGMR, 2005): (1) The block based tagging, and (2) The entropy based tagging (or called non-block based tagging).

With the block based tagging, we need to define the haplotype block first. Inside each haplotype block, the SNPs are in strong LD with each other, while, for SNPs of different blocks, they are of low LD. The disadvantage of this type of tagging SNPs is that the definition of the haplotype block is not unique and sometimes ambiguous, as we will see later. Also, it is true that the coverage of the haplotype block is not enough in some genomic region.

Because of the problems associated with the haplotype blocks, alternative methods have been developed and they can collectively called entropy based tagging. The term entropy is used loosely as the measure for assessing the amount of information that can be captured or represented by these tag SNPs. In this approach, it is not necessary to define the haplotypes and then to define the haplotype blocks.

Instead, the goal of this approach is to select a subset of SNPs (the tag SNPs) that can capture the most information across the genomic region. Different multivariate statistical techniques have been applied to achieve this task. Byng et al. (2003) proposed the use of single and complete linkage hierarchical cluster analysis to select tag SNPs. Hierarchical clustering starts with a square matrix of pair-wise distances between the objects to be clustered. For the problem of tag SNP selection, the objects to be clustered are the SNPs, and an appropriate measure of distance is $1 - R^2$, where R^2 is the squared correlation between two SNPs. The rationale is this: the required sample size for a tag SNP to detect an indirect association with a disease is inversely proportional to the R^2 between the tag SNP and the causal SNP.

4.1.3.2 Existing Programs for Tag-SNP Selection

There exist different software packages for the tag SNP selection. In the official web page of The Centre for Integrated Genomic Medical Research (CIGMR, 2005) of The University of Manchester, U.K., the following programs are listed with the program web addresses:

CLUSTAG: http://hkumath.hku.hk/web/link/CLUSTAG/CLUSTAG.html
HapBlock: http://www.cmb.usc.edu/msms/HapBlock/
htSNP: http://www-gene.cimr.cam.ac.uk/clayton/software/stata/htSNP/
htSNPer: http://www.chgb.org.cn/htSNPer/htSNPer.htm
SNPtagger: http://www.well.ox.ac.uk/~xiayi/haplotype/
TAG 'n' TELL: http://snp.cgb.ki.se/tagntell/

Many of the above programs have been tested with samples of size about 100 SNPs only. As an example, the TAG 'n' TELL v2.0 (the current version as on 16th April 2005) has stated on its web that it can handle with the number of markers up to 30. Besides our program CLUSTAG, all the other programs are based on the concept of haplotype block. Most of the software packages do not have the graphical outputs for displaying the tagging information, while the CLUSTAG has its graphical output in the hyper-text format that enables the users to display their CLUSTAG results on the web easily (Fig. 4.1).

The HapBlock program (Zhang et al., 2005) calculates the haplotype frequencies with the PL-EM algorithm (Qin et al., 2002). The PL-EM algorithm is performed for each set of consecutive SNPs so that they can from a potential block, rather than for the whole set of SNPs. The program has the options to choose three of the many definitions of the haplotype blocks and these three block definitions are basing on:

1. Common haplotypes (Patil et al., 2001; Zhang et al., 2002)
2. The LD measure with D' (Gabriel et al., 2002) and
3. The four-gamete test (Wang et al., 2002)

The selection of haplotype tagging SNPs are based the power of these SNPs for distinguishing haplotypes. Another criterion is based on the haplotype diversity (Clayton 2001). The proportion of haplotype diversity is computed for a subset of

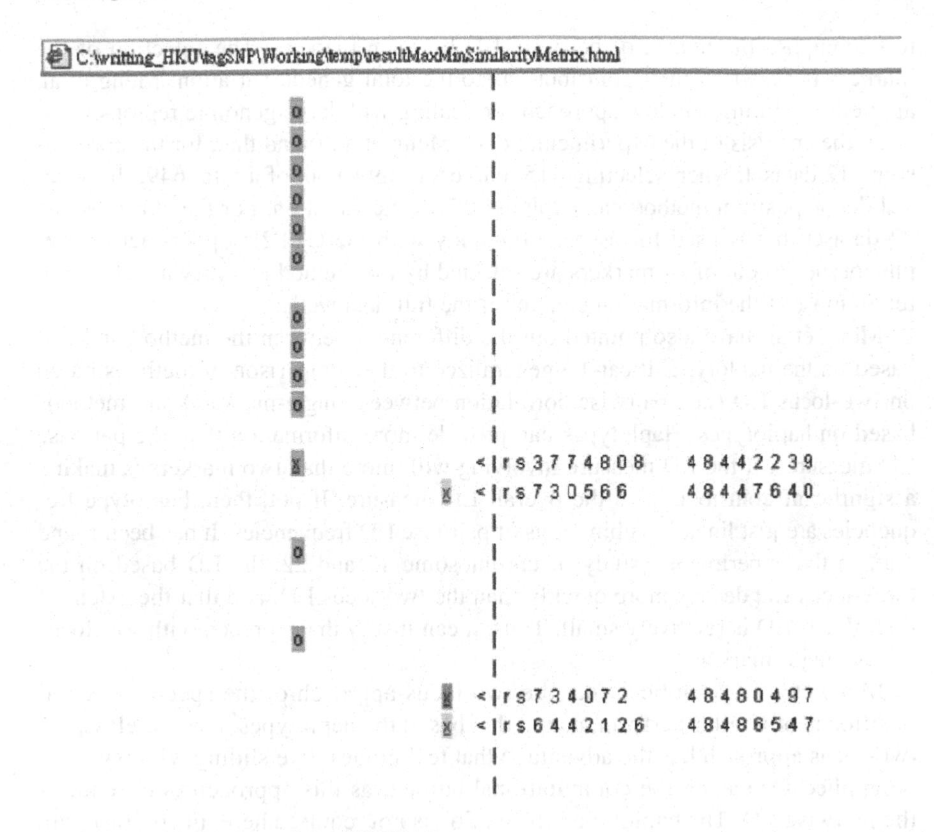

Fig. 4.1 Graphical output of the CLUSTAG

SNPs over all SNPs in each block. The minimum set of SNPs are chosen with the limitation that at least a pre-defined percent of overall haplotype diversity can be explained by these chosen subset. Similarly, the programs htSNP, SNPtagger, and Tag 'n' TELL are also basing on this haplotype diversity. The program htSNPer also relies on firstly defining the haplotype blocks and then computes the haplotype tagging SNPs. The users can select four common methods to define the blocks. The program can select the haplotype tagging SNPs that cover most haplotypes within the block or that have the largest haplotype diversity. An advantage of the program is that it employs the Branch-and-Bound (BB) algorithm for accelerating the searching the haplotype tagging SNPs.

4.1.3.3 Motivations for Developing Non-block Based Tagging Methods

Meng et al. (2003) noticed that all the block-detecting methods can result in different block boundaries. In fact, the existence of the block is still conflicting (Couzin, 2002). As a result, Meng et al. proposed a method that uses the spectral decomposition to decompose the matrix of pairwise LD between markers. The selection of the markers is based on their contributions to the total genetic variation. Meng et al. applied the sliding window approach for dealing with large genomic regions.

In the analysis of the experimental data, Meng et al. found that, for the chromosome 12 dataset, when selecting 415 markers (63.9%) out of a total 649, the spectral decomposition method can explain 90% of the variation. For the chromosome 22 dataset that is used for association study with the CYP2D6 poor-metabolizer phenotype, 20 out of 27 markers are selected by the method and they are shown to retain most of the information content of the full data well.

Meng et al. have also pointed out the differences between the method and that based on the haplotype. It can be generalized to the comparison of methods based on two-locus LD (i.e., pairwise correlation between single-markers) and methods based on haplotypes. Haplotypes can provide more information than the pairwise LD measures, if the LD measure involving with more than two markers is making a significant contribution to the overall LD measure. If not, then, haplotype frequencies are just linear combinations of pairwise LD frequencies. It has been found that, in the experimental study of chromosome 12 and 22, the LD based on the three-locus LD decays more quickly than the two-locus LD, and that the extent of three-locus LD is relatively small. Thus, it can justify the approach with two-locus LD of single-markers.

Meng et al. also noticed that the two-locus approach of the spectral decomposition is of similar performance to that based the haplotypes. Nevertheless, the two-locus approach has the advantage that techniques, like sliding windows, can be applied for easing the computational burden, as this approach only requires the pairwise LD. The haplotype information is not required here, in contrast with the haplotype-based method, which requires the estimation of the haplotype frequencies with the numerical methods like EM algorithms. The computational time for such algorithms will increase dramatically when the number of markers increases.

4.2 CLUSTAG: Its Theory

Our program CLUSTAG has been developed with the clustering algorithms of the multivariate statistics. A desirable property for a clustering algorithm for the tag-SNP selection would be that a cluster must contain at least one SNP (the tag SNP) that is no more than the merging distance from all the other SNPs from the same cluster. If this is the case, then by setting a cutoff merging distance of C, one can ensure that no SNP is further than C away from the tag SNP in its cluster. In this sense, neither of the methods proposed by Byng et al. (2003) is ideal, since the single-linkage method does not guarantee the existence of a tag SNP with distance less than C from all SNPs in the same cluster, while complete-linkage is too conservative in that all SNPs have distance under C from all other SNPs in the same cluster.

4.2.1 *Definition of the Clustering Process*

In an m-clustering of a data set ***X***, it is to group the partition of ***X***, ***R***, into m sets (clusters), $C_1, \ldots, C_m$, such that the below conditions are satisfied (Theodoridis and Koutroumbas, 2003):

1. $C_i \neq \emptyset,\ i = 1,\ldots,m$
2. $\bigcup_{i=1}^{m} C_i = X$
3. $C_i \cap C_j = \emptyset,\ i \neq j,\ i, j = 1,\ldots,m$

It is also required that the vectors inside a cluster C_i are more similar to each other, and less similar to the feature vectors of other clusters.

4.2.2 *Similarity Measures*

Similarity measures are needed to give us a quantitative comparison of how close the vectors are in the clustering process. A similarity measure (SM) s on ***X*** is defined as:

$$s : X \times X \rightarrow \Re$$

such that it satisfies:

$$\exists s_0 \in \Re : -\infty < s(x, y) \leq s_0 < +\infty, \forall x, y \in X$$

and also

$$s(x, x) = s_0, \forall x \in X$$
$$s(x, y) = s(y, x), \forall x, y \in X$$

It should be noted that, if the similarity measure also satisfies the following two conditions, then it would be a metric SM (Theodoridis and Koutroumbas, 2003):

$$s(x,y)=s_0 \text{ if and only if } x = y$$

and

$$s(x, y)s(y, z) \leq [s(x, y) + s(y, z)]s(x, z), \forall x, y, z \in X.$$

The distance (dissimilarity) measure d can be defined similarly, with:

$$d : X \times X \rightarrow \Re$$

satisfying the conditions:

$$\exists d_0 \in \Re : -\infty < d_0 \leq d(x, y) < +\infty, \forall x, y \in X$$

and

$$d(x,x) = d_0, \forall x \in X$$

and

$$d(x,y) = d(y,x), \forall x,y \in X$$

The above similarity measure and distance measure is for members within a set. Now, we are going to define the similarity measure between sets of data points. We will see later that hierarchical clustering utilizes such a measure in its clustering process.

Mathematically, let S^{ss} be the similarity function between different sets, and let X_i, X_j denote two sets of data vectors. The common similarity functions are:

1. The *max* similarity function:

$$S^{ss}_{\max}(X_i, X_j) = \max_{x \in X_i, y \in X_j} s(x,y)$$

where s is a similarity measure between two vectors.

2. The *min* similarity function:

$$S^{ss}_{\min}(X_i, X_j) = \min_{x \in X_i, y \in X_j} s(x,y)$$

3. The *average* similarity function:

$$S^{ss}_{avg}(X_i, X_j) = \frac{1}{n_{X_i} n_{X_j}} \sum_{x \in X_i} \sum_{x \in X_j} S(x,y)$$

where n_{Xi} and n_{Xj} are the cardinalities of the sets X_i and X_j respectively.

It can be noticed that the above definitions of the similarity functions are based on the similarity measures between two vectors. With the different definitions of the similarity measures between vectors, we can usually obtain different clustering results. Thus, by choosing the similarity measure carefully, one can have clustering results that fit a particular objective better. There have different works of applying the different clustering algorithms on the genome-wide microarray studies (for example, Mcshane et al., 2002; Datta and Datta, 2003). Later on, we will see how our proposed definition of minimax function for the tag-SNP selection can improve the clustering results of the single nucleotide polymorphisms.

4.2.3 Agglomerative Clustering

The clustering algorithms of the CLUSTAG are of agglomerative clustering, where the two clusters with the smallest inter-cluster distance are successively merged until all the objects have been merged into a single cluster. Different forms of

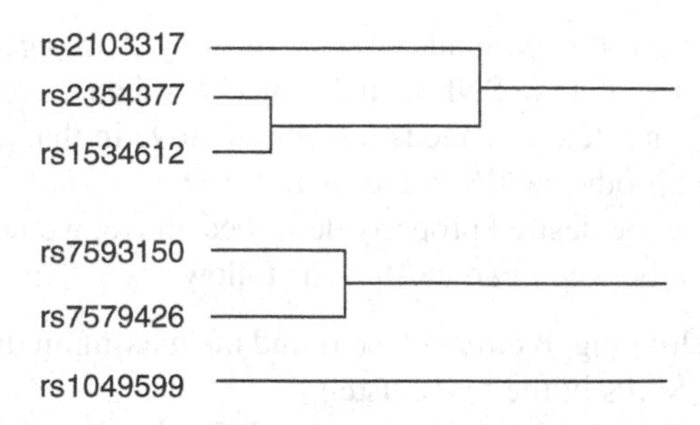

Fig. 4.2 Sample illustrative dendrogram showing how seven SNPs are merged into three clusters at or below the cutoff merging distance

agglomerative clustering differ in the definition of the distance between two clusters, each of which may contain more than one object. In single-linkage or nearest-neighbour clustering, the distance between two clusters is the distance between the nearest pair of objects, one from each cluster. In complete linkage or farthest neighbour clustering, the distance between two clusters is the distance between the farthest pair of objects, one from each cluster. The clustering process can be represented by a dendrogram. The dendrogram can show how the individual objects are successively merged at greater distances into larger and fewer clusters. All distinct clusters that have been generated at or below a certain user-defined distance are considered (see Fig. 4.2). In this example of complete linkage clustering, the distances between rs2103317, rs2354377 and rs1534612 are less than the user-defined distance. So are the distances between rs7593150 and rs7579426.

The agglomerative algorithms can be described with the concept of nesting (Theodoridis and Koutroumbas, 2003). Let $\Re_1$ of k clusters and $\Re_2$ of r clusters be two clusters form in the clustering process, where $r < k$. When each cluster in $\Re_1$ is a subset of a set in $\Re_2$, then we say that $\Re_1$ is nested in $\Re_2$ and we denote this by $\Re_1 \subset \Re_2$. For example, let $\Re_1 = \{ \{x_2, x_4\}, \{x_1\}, \{x_3, x_5\} \}$ and $\Re_2 = \{ \{x_2, x_4\}, \{x_1, x_3, x_5\} \}$, then we say that $\Re_1$ is nested in $\Re_2$. The hierarchy of the agglomerative algorithms is as follow. Let the initial clustering be ℜ$_{0}$, the first clustering be $\Re_1$,..., and the final clustering be $\Re_{N-1}$. The hierarchy is $\Re_0 \subset \Re_1 \subset \ldots \subset \Re_{N-1}$.

4.2.4 *Clustering Algorithm with Minimax for Measuring Distances Between Clusters, and Graph Algorithm*

A desirable property for a clustering algorithm, in the context of tag-SNP selection, would be that a cluster must contain at least one SNP (the tag SNP) that is no more than the merging distance from all the other SNPs from the same cluster. If this is the case, then by setting a cutoff merging distance of C, one can ensure that no SNP is further than C away from the tag SNP in its cluster. As said, neither of the methods proposed by Byng et al. (2003) is ideal, since the single-linkage method does not guarantee the existence of a tag SNP with distance less than C from all SNPs in the same cluster, while complete-linkage is too conservative in that all SNPs have distance under C from all other SNPs in the same cluster.

In order to achieve the desired property described above, we propose a new definition of the distance between two clusters, as follows:

1. For each SNP belonging to either cluster, find the maximum distance between it and all the other SNPs in the two clusters.
2. The smallest of these maximum distances is defined as the distance between the two clusters.
3. The corresponding SNP is defined as the tag SNP of the newly merged cluster.

We call this method minimax clustering, which is an agglomerative method. There is a parallel in topology in which the distance between two compact sets can be measured by a sup-inf metric known as Hausdorff distance (Barnsley, 1988; Wucklidge, 1996).

For comparison we have also implemented an algorithm based on the NP-complete minimum dominating set of the set-cover problem in the graph theory, similar to the greedy algorithm developed by Carlson et al. (2004). The set of SNPs are the nodes of a graph, which are connected by edges where their corresponding SNPs have $R^2 > C$. The objective is to find a subset of nodes such that that all nodes are connected directly to at least one SNP of that subset. The details of this heuristic algorithm can be found in Reuven and Zehavit (2004), Fujito (2001) and Johnson (1973). The one by Johnson (1973) is on the studies of the error bound of the algorithm and the one by Fujito (2001) studies the case of weighted edges. Briefly, at the beginning of the method, all the SNPs belong to the untagged set. The algorithm picks the node with the largest number of nodes that are connected directly to it (without passing through any other nodes) from the untagged set. Then the SNPs inside the selected subset are deleted from the untagged set, and the next largest connected subset is chosen from the untagged set. The algorithm terminates when the untagged set becomes empty.

4.3 Experimental Results of CLUSTAG

4.3.1 Experimental Results of CLUSTAG and Results Comparisons

We implemented the complete linkage, minimax linkage and set cover algorithms in the program CLUSTAG. The program takes a file of R^2 values produced, for example, by HAPLOVIEW (Barrett et al., 2005), and outputs a text file containing one row per SNP and the following columns (Fig. 4.3): (i) SNP name, (ii) cluster number, (iii) chromosomal position, (iv) minor allele frequency, (v) maximal distance

SNP NAME	Cluster	Tagged	SNP Positions	MAF	Max(1-R^2)	Order by Max(1-R^2)	Average(1-R^2)	Order by Average(1-R^2)
rs2305634	0	0	47004130	0.36	0.19	20	0.091666676	19
rs1079276	0	0	47006331	0.45	0.14999998	9	0.076666676	16
rs7646799	0	0	47113159	0.45	0.14999998	9	0.077916674	18
rs11130115	0	0	47142253	0.45	0.14999998	9	0.076666676	16
rs2305638	0	0	47007434	0.46	0.18	17	0.09625002	21
rs7610636	0	0	47025028	0.47	0.19	20	0.102500014	23
rs4078466	0	0	47087135	0.46	0.18	17	0.09583335	20
rs6785790	0	0	47107524	0.46	0.19	20	0.097916685	22
rs4315703	0	0	47283373	0.42	0.13999999	3	0.04833333	4
rs295441	0	0	47293887	0.43	0.13	1	0.047916662	2
rs4683327	0	0	47319854	0.42	0.13999999	3	0.04833333	4
rs2159400	0	0	47336482	0.43	0.13	1	0.047916662	2
rs295458	0	0	47346177	0.42	0.13999999	3	0.04833333	4
rs807931	0	0	47349909	0.4	0.14999998	9	0.05166666	9
rs7613282	0	0	47350001	0.42	0.13999999	3	0.04833333	4
rs11130127	0	0	47359385	0.39	0.13999999	3	0.049999993	8
rs922957	0	1	47382744	0.42	0.13999999	3	0.040000003	1
rs2062278	0	0	47377353	0.42	0.17000002	13	0.060416657	10
rs10865946	0	0	47385652	0.41	0.17000002	13	0.064583324	13
rs11712445	0	0	47395144	0.41	0.19	20	0.070833325	15
rs4858811	0	0	47441116	0.42	0.17000002	13	0.060416657	10
rs11130128	0	0	47447224	0.42	0.17000002	13	0.060416657	10
rs2101247	0	0	47454895	0.4	0.18	17	0.06624999	14
rs6442055	1	0	47071314	0.43	0.13999999	1	0.051999997	2
rs6766230	1	1	47156801	0.42	0.13999999	1	0.043999992	1

Fig. 4.3 Text output of the CLUSTAG

(1 - R^2) from other SNPs in the same cluster, and (vi) average distance (1 - R^2) from other SNPs in the cluster. Both (v) and (vi) are useful for providing alternative SNPs that can serve as the tag SNP of the cluster, allowing some flexibility in the construction of multiplex SNP assays. A visual display (in html format) provides a representation of the SNPs in their chromosomal locations, color-labeled to indicate cluster membership (Fig. 4.1). The tag SNP is highlighted and hyperlinked to a text box containing columns (i)–(vi) on the cluster.

We have compared the performance of the three implemented algorithms, using SNP data from the ENCODE regions of the HapMap project, according to three criteria:

1. Compression, the ratio of clusters to SNPs
2. Compactness, the average distance between a SNP and the tag SNP of its cluster (1 - R^2), and
3. Run time

Our results show that the compression ratio is roughly equivalent for the set cover and minimax clustering algorithms but substantially higher for the complete linkage (Table 4.1). The minimax algorithm produces more compact clusters than the set cover algorithm (Table 4.2), but takes approximately twice as long to run. The run times of all three algorithms are expected to increase in proportion to the square of the number of SNPs.

The complexity of the clustering methods are of order $O(n^2)$. With the run time information in our table of several hundred SNPs and this complexity information,

Table 4.1 Properties of three tag SNP selection algorithms, evaluated for ENCODE regions

Encode Region	Compression			Run time (s)		
(SNP no.)	Complete	Minimax	Set cover	Complete	Minimax	Set cover
2A (519)	0.277	0.245	0.247	3.94	5.42	3.20
2B (595)	0.291	0.255	0.261	5.44	6.92	4.03
4 (665)	0.242	0.211	0.209	6.53	13.30	5.25
7A (417)	0.314	0.281	0.281	2.56	3.39	2.00
7B (463)	0.186	0.166	0.171	3.53	5.03	2.84
7C (433)	0.240	0.217	0.215	2.38	3.28	1.80
8A (364)	0.269	0.245	0.245	2.39	2.94	1.83
9 (258)	0.360	0.318	0.314	1.47	1.74	0.98
12 (454)	0.260	0.227	0.227	2.69	3.69	2.03
18 (350)	0.283	0.254	0.254	2.17	2.81	1.64

Table 4.2 Compactness of three tag SNP selection algorithms, evaluated for ENCODE regions

Encode region	Compactness		
(SNP no.)	Complete	Minimax	Set cover
2A (519)	0.021	0.033	0.037
2B (595)	0.018	0.033	0.032
4 (665)	0.016	0.031	0.035
7A (417)	0.013	0.028	0.032
7B (463)	0.020	0.030	0.035
7C (433)	0.018	0.019	0.021
8A (364)	0.019	0.035	0.040
9 (258)	0.012	0.025	0.031
12 (454)	0.017	0.028	0.034
18 (350)	0.014	0.033	0.037

the users can estimate roughly the expected run time for their samples before the program's execution. The run time will not be an issue for data of several hundred to a hundred thousand SNPs. But, it will be a constraint when we are studying the whole genome at one time, when the size may be of several million SNPs. This is an area of further work as the HAPMAP project is producing the whole genome haplotype information.

We have also tested the different threshold values C for the chromosome 9 of the ENCODE data in the following two figures (Figs. 4.4 and 4.5). The values of the threshold C are 0.7, 0.75, 0.8, 0.85, 0.9 and 0.95, which cover the range of reasonable threshold values. The results show that the compression ratio and the compactness are quite stable over the range from 0.7 to 0.8.

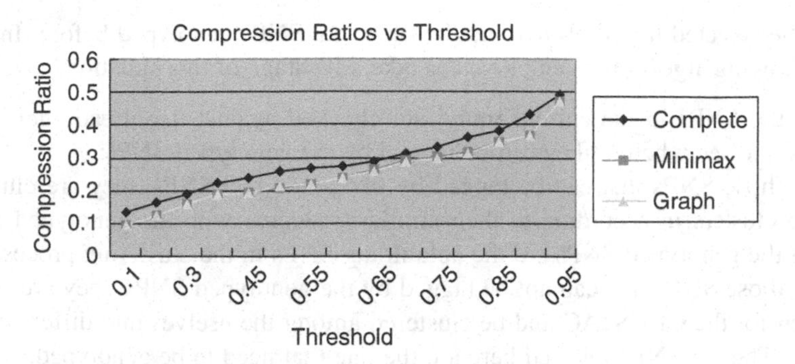

Fig. 4.4 Compression ratios vs. different threshold values

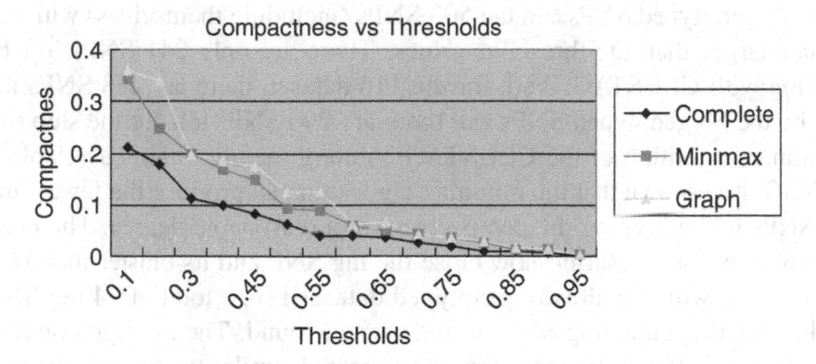

Fig. 4.5 Compactness vs. different threshold values

4.3.2 Practical Medical Case Study with CLUSTAG

We applied our tailor-made data mining algorithm to a medical problem in the HKU's Genome Research Centre. There is a set of SNPs that are going to be clustered by CLUSTAG. The CLUSTAG is employed for selecting the tag SNPs with the threshold at 0.8. Prior to the tagging step, there are two set of tag SNPs that have already been genotyped. The objective is to use the information provided by the previously genotyped SNPs in order to save the number of tag SNPs in this new round of genotyping.

The input files for the problem include the LD file of between all the SNP pairs, the SNP map file containing the SNP position and MAF information, and the two set of tag SNP that have been genotyped before (called it dataset 116 and dataset 144). The total number of SNPs that are going to study in the current round is 816. In the dataset 144, there exist 72 tag SNP data that can be found be in the current round. In the dataset 116, that number is 58 tag SNPs.

In this SNP problem, there exists some SNP genotyping information that has been obtained through previous experiments. If the CLUSTAG is applied to all the data, the selected tag SNPs may not be those tag SNPs genotyped before. Instead, the following algorithm is employed to take advantage of this situation:

- All the SNPs in the current round are checked against the already-genotyped SNPs to see whether they can be tagged by the genotyped SNPs;
- For those SNPs that can be tagged by the genotyped SNPs, they are clustered into clusters in according to their similarity degree with the genotyped SNPs, and the genotyped SNPs are the default tag SNPs in the clustering process;
- For those SNPs that can not be tagged by the genotyped SNPs, they are formulated for the CLUSTAG and be clustered among themselves into different clusters. The tag SNPs selected here are the one that need to be genotyped.

It is found that the steps (1) and (2) can effectively reduce the number of SNPs that are needed for the clustering process to select tag SNP in step (3). For the dataset 144, the 72 genotyped SNPs can tag 503 SNPs (including themselves) with similarity values larger than the threshold values. There are only 241 SNPs left for the processing with CLUSTAG. And, for the 116 dataset, there are 468 SNPs that are tagged by the 58 genotyped SNPs and there are 290 SNPs left for the step (3) with the minimax algorithm of the CLUSTAG. Among the algorithms available in the CLUSTAG, it is shown that the minimax algorithm can produce the fewest number of tag SNPs while keeping the compactness at a reasonable degree. The compactness number is for measuring how close the tag SNP and its cluster members are. In the step (3), with the already-genotyped dataset 116, a total of 84 tag SNPs are needed to tag the remaining SNPs in the current round. The average compactness for the clusters is 0.014. It means that the expected similarity between the tag SNP and its cluster members are 0.986. The running time for the step (3) is 2.297 s. For the dataset 144, a total of 74 SNPs are chosen as tag SNPs. The average compactness is 0.0112 and the running time is 1.891 s.

It should be noted that the compactness for the clusters in the step (2) is generally of higher value than that in step (3). It is because, in the clustering process in step (2), the tag SNPs are by default chosen as those of the already-genotyped SNPs. For example, the average compactness for the clusters of dataset 116 is 0.053, and that with the dataset 144 is 0.050.

In another genomic experiment, Brookes et al. (2006) have investigated the attention deficit hyperactivity disorder (ADHD) with the utilization of the CLUSTAG for tag-SNP selection. ADHD is a common neurodevelopmental disorder, which starts in early childhood and persists into adulthood in the majority of cases. The SNP genotyping experiment was by the Illumina high-throughput BeadArray technology. Data obtained with Illumina included a list of SNPs and their estimated genotype success rate for each gene. For tag SNPs selected from HapMap, it was fortunate that there often existed additional SNPs that were within a SNP cluster, and SNP with high estimate of success genotype rate could be selected in preference to one with a low rate.

Brookes et al. screened for association of the disease with the common genetic variations in the target regions. The regions are identified with the biological knowledge of the genes through their effect of regulation of dopamine, serotonin and norepinephrine neurotransmission. CLUSTAG algorithm has been applied in this study for selecting the tag SNPs. In order to ensure a high chance of detecting indirect association, a threshold for R^2 of larger than 0.8 is used. The second method used in this study for tag-SNP selection is the newer algorithm in Haploview. The tag SNP recommendation by the CLUSTAG algorithm is preferentially selected, when the two methods recommended selection of two different SNPs that fell within the same cluster defined by CLUSTAG. This can avoid redundancy of the marker information. The details are available in their study. Through this gene-wide test, the disease was identified to have associations with TPH2, ARRB2, SYP, DAT1, ADRB2, HES1, MAOA and PNMT for further studies.

4.4 WCLUSTAG: Its Theory and Application for Functional and Linkage Disequilibrium Information

4.4.1 Motivations for Combining Functional and Linkage Disequilibrium Information in the Tag-SNP Selection

In the association studies for complex diseases, there are mainly two approaches for selecting the candidate polymorphisms. In the functional approach, the candidate polymorphisms are selected if they are found to cause a change in the amino acid sequence or gene expressions. The second approach, the positional approach, is to systematically screen polymorphisms in a particular genome region by using the linkage disequilibrium information with the disease-related functional variants. The functional approach is direct approach, while the positional approach is indirect approach. The algorithms and programs that we have described in the above sections are basically constructed with the positional approach. The candidate tag SNPs are selected for genotyping by utilizing the redundancy between near-by SNPs through the LD information. The purpose is to improve the efficiency of the analysis with minimal loss of information while reducing the genotyping costs at the same time.

In order to further utilize the genomic information for improving the tag-SNP selection efficiency, it would be desirable if the tag-SNP selection algorithm can take account of the functional information, as well as the LD information. In the human genome, it is well known that different kinds of polymorphisms have different effects on the gene expressions and importance. The SNPs can attach more importance when their positions are within the coding regulatory regions. Similarly, for SNPs in the non-coding regions, they are attached with less biological importance. Furthermore, it is also desirable for the tag-SNP selection algorithm to take care of practical laboratory considerations like the readiness of the SNPs for assaying and the existing genotyped results in the previous laboratory experiments.

4.4.2 Constructions of the Asymmetric Distance Matrix for Clustering

The WCLUSTAG is developed in order to take care of the functional information and LD information, as well as the laboratory consideration. The development of the WCLUSTAG is based on the previous CLUSTAG, by adding the variable tagging threshold and other functions, and the web-based interface. As described above, the CLUSTAG is of agglomerative hierarchical clustering and starts with the constructing of a square matrix of pair-wise distance between the objects to be clustered. An appropriate distance measure for the LD tagging is $1 - R^2$, where the second term is the square of the correlation between the SNPs. The clusters with the least inter-cluster distance are successively merged with each other. A cutoff merging distance, denoted by C, is required for the terminating of the algorithm and for ensuring that, in each cluster, it contains no SNP further than C away from the corresponding tag SNP.

In CLUSTAG, this cutoff merging distance C is the same for all the SNPs. In WCLUSTAG, the program has been modified so that the tagging threshold C can be specified by the user for each SNP, and can be different among the SNPs. Then, the factors like the positional and functional information, as well as the practical laboratory information, can be utilized for the assigning of tagging threshold values for SNPs individually. For example, for SNPs in the coding or regulatory regions, a high value of C (e.g. 0.8) can be assigned to these SNPs. On the other hand, for the other SNPs in the non-coding regions etc., a low value of C (like 0.4) can be given. With this modification, unlike the CLUSTAG, the square matrix of pair-wise distances between the objects becomes asymmetric for WCLUSTAG. For example, let a coding SNP have a C of 0.8, and another non-coding SNP of C value 0.4, and let the R^2 between these two SNPs be 0.5. It can be observed that the first SNP can serve as the tag SNP for the second. On the hand, the second SNP is not able to tag the first one. Thus, the WCLUSTAG has been built with the capability for handling of asymmetric distance matrix, such that the distance from object h to object k is not required to be the same as the distance from object k to object h.

With these considerations, the WCLUSTAG has been modified from CLUSTAG and works as followed:

- Firstly, a user-define value C is assigned for each SNP;
- Secondly, let C_k be the value of C for SNP k, and, let the distance from SNP h to SNP k be $C_k - R_{hk}{}^2$. Then, for $C_k - R_{hk}{}^2 < 0$, SNP h can serve as a tag SNP for SNP k.
- Thirdly, the minimax clustering method is applied with this new asymmetric distance matrix, and the cut-off merging distance is zero.

Then, cluster is formed for the case that there is a tag SNP that has a distance of zero or less with its cluster members respectively. The set-cover algorithm has undergone similar modifications in WCLUSTAG.

4.4.3 Handling of the Additional Genomic Information

As discussed above, it is desirable that the tag-selection algorithm can initially select all SNPs that have already been genotyped, and then remove these SNPs and the SNPs tagged by these SNPs from the next genotyping experiment. The algorithm will provide the laboratory users with more flexibility if the algorithm can exclude those SNPs that have problems with assay design etc. In order to achieve these properties, the algorithm has been subjected to the following modifications, which can be done by changing the values of certain elements in the matrix similarities $[R^2_{hk}]$.

For the case that the SNP t has already been genotyped, all the elements of column t in the matrix are set to zeros, except for the diagonal element of the column t which remains one. This setting can ensure that the SNP t can not be tagged by any other SNPs, and, therefore, it will be included as one of the tag SNPs in the clustering and graph algorithms. For the case that the SNP t has problem with assay design, all the elements of the row t in the matrix are set to zero. Therefore, the SNP t can never serve as one of the tag SNPs in the algorithms. There is one problem associated with these settings. With these settings, it does not ensure that all the problematic SNPs for assay design can be tagged in the algorithms. This is because some non-assayable SNPs can only be tagged by certain SNPs, while these SNPs may not be selected as the tag SNPs with the algorithms. This problem can be solved with the following further modification, which forces the selection of certain SNPs for tagging these non-assayable SNPs.

Firstly, for non-assayable SNPs that can not be tagged by any assayable SNP, as there does not exist any assayable tag SNP for them, they are listed and excluded from further processing. Then, the remaining non-assayable SNPs are subjected to following procedure to ensure that there will exist at least one tag SNP for each of them:

1. The set of already-genotyped SNPs (if existed) are checked if the SNPs there can tag the non-assayable SNPs. The SNPs of the non-assayable SNPs that can not be tagged by these already-genotyped SNPs are called the set of untagged non-assayable SNPs.
2. Each assayable SNPs (but not those already genotyped) is checked against the untagged non-assayable SNPs for the number of untagged non-assayable SNPs that each assayable SNP can tag. The one with the largest number is assigned as a SNP for forced selection, and the non-assayable SNPs that can be tagged by this SNP are removed from the set of untagged non-assayable SNPs.
3. For cases that there still exist untagged non-assayable SNPs, the above step (2) is repeated until there is no untagged non-assayable SNP.

The SNPs selected in the above steps (2) and (3) are treated in the same way as the SNPs that have been already genotyped, and are subjected to the same procedure for forced selection.

4.5 WCLUSTAG Experimental Genomic Results

To illustrate the performance of the new algorithms, the CEPH sample genotype data from the International Haplotype Map Project was tested with the algorithms. The ENCODE regions were selected because genotyping were undertaken for all known SNPs in these regions. Intragenic regions were identified from the start and end points of the coding sequences for the 33K Ensembl genes in NCBI build 34. Intragenic SNPs are given a C weighting of 0.8, and other SNPs 0.4. The compression ratios (the number of tag-SNPs over the total number of SNPs) of the various ENCODE regions are compared with the original procedure which used a uniform C value of 0.8. Our results show that there can be a further 35.2% saving with our weighted minimax algorithm, and 35.9% with the set cover method (Table 4.3). We also explored the impact of using different weighting schemes. Some additional saving can be obtaining by lowering the weights for either intragenic or other SNPs, although the compression ratios remain in the region of 0.2 (Table 4.4). The average ratio of the SNPs in the intragenic regions to the overall SNPs is 32.3% in the dataset (Table 4.5).

Table 4.3 Properties of the tag SNP selection algorithms, weighted with 0.8 for gene regions and 0.4 for other regions

Encode region	Compression (Uniform)			Compression (Weighted)	
(SNP no.)	Complete	Minimax	Set cover	Minimax	Set cover
2A (519)	0.277	0.245	0.247	0.104	0.104
2B (595)	0.291	0.255	0.261	0.197	0.198
4 (665)	0.242	0.211	0.209	0.089	0.089
7A (417)	0.314	0.281	0.281	0.149	0.139
7B (463)	0.186	0.166	0.171	0.114	0.114
7C (433)	0.240	0.217	0.215	0.189	0.185
8A (364)	0.269	0.245	0.245	0.190	0.190
9 (258)	0.360	0.318	0.314	0.221	0.225
12 (454)	0.260	0.227	0.227	0.167	0.163
18 (350)	0.283	0.254	0.254	0.186	0.189
Average	0,267	2.237	0.238	0.154	0.153
Additional saving	–	–	–	35.2%	35.9%

Table 4.4 Effect of weighting scheme (intragenic versus other SNPs) on the comparison ratios for tag-SNP selection algorithms in the Chromosome 9 Encode data

	Compression	
Weighted ratio	Minimax	Set cover
0.8:0.4	0.221	0.225
0.8:0.3	0.198	0.198
0.8:0.5	0.240	0.244
0.7:0.4	0.217	0.221
0.9:0.4	0.240	0.244
0.8:0.8	0.318	0.314

Table 4.5 The number of SNPs in the intragenic regions and the other regions. The average ratio of the SNPs in the intragenic regions to the overall SNPs is 32.3%

SNPs no.	SNPs in intragenic regions	SNPs in other regions
chr2A	0	519
chr2B	273	322
chr4	0	665
chr7A	21	396
chr7B	159	304
chr7C	299	134
chr8A	203	161
chr9	66	192
chr12	180	274
chr18	167	183

4.6 Result Discussions

With the necessary modifications, the WCLUSTAG can enable the users to select tag SNPs, with the advantage of both the functional approach and the positional approach. The choice of the threshold values can be made according to the budget for the disease data. Currently, the users can use the downloadable program version, which may be convenient for running scripts for multiple data sets. Or, the users can assess our web interface for importing their own genotype data. The web interface also has the capability of downloading the HapMap data directly from its mirror database for further computation.

There are factors that can affect the overall effectiveness of the tagging strategy. They include the functional information like the comprehensiveness of SNP maps, the quality of functional annotation of the genome, and the linkage disequilibrium information between the polymorphisms and the complex human diseases, and the underlying genetic architecture of the complex diseases. Many of these have not been fully understood by researchers and remain to be explored in the future studies.

Chapter 5
Case Study II: Constrained Unidimensional Scaling for Linkage Disequilibrium Maps

With the advance of the genotyping single nucleotide polymorphisms (SNPs) in mass scale of high density in a candidate region of the human genome, the linkage disequilibrium analysis can offer a much higher resolution of the biological samples than the traditional linkage maps. The advantages of LD maps include the revealing of the fine scale recombination patterns, the facilitating of the optimal SNP/marker spacing, and the increasing of the power for localizing disease genes etc. The first LD maps were proposed by Maniatis and colleagues (Maniatis et al., 2002). The derivation of this LD map is parametric and requires the estimation of three coefficient parameters. Nevertheless, these estimated parameters are found to have large variances among different populations.

We have formulated this LD mapping problem as a constrained unidimensional scaling problem. Our method, which is directly based on the measurement of LD among SNPs, is non-parametric. Therefore it is different from LD maps derived from the given Malecot model. We have proposed the quadratic programming approach for solving this constrained unidimensional scaling problem. Different from the classical metric unidimensional scaling problem, the constrained problem is not an NP-hard combinatorial problem. The optimal solution is determined by using the quadratic programming solver.

5.1 Background

5.1.1 Linkage Analysis and Association Studies

5.1.1.1 Linkage Analysis

The linkage analysis has been using for the identification of the Mendelian diseases (Barnes and Gray, 2003). It can find out the correlation between the phenotypic patterns (like the disease state) and the genetic markers. The analysis is based on the fact that there are only a small number of recombination in the two to five generation families that are used for linkage analysis. Recombination is the process that the chromosome undergoes breakage and exchanges segments of DNA during meiosis, whereby gametes (eggs and sperm) are produced (Alberts et al., 1994). These markers and diseases correlations can extend over many megabases (Mb). A linkage scanning of as few as several hundred simple tandem repeat (STR) etc., can give us enough coverage of the entire human genome.

Chromatids refer to the sister strands duplicated from the chromosomes during the early stage of meiosis (Sham, 1998). The resulting complex during the meiosis is called a tetrad. The region where the non-sister chromatids attach to each other is called chiasmata. The crossing over chromatids can be found in these chiasmata. From the observations of the meioses process, it is suggested that there is at least one chiasma in each chromosome. The result is that there is an obligatory crossover.

The genetic map distance between two loci is measured in units of Morgans. It is defined as the expected number of crossovers between the loci on a single chromatid during meiosis. In the meiosis process, each tetrad of the chromosome has four chromatids and each crossover involves two chromatids, so the genetic distance is also equal to half of the number of crossover in the tetrad. A smaller unit of the Morgans is the centiMorgan (cM), which is one hundred times smaller than a Morgan. Roughly, one centiMorgan corresponds to 1,000 kB.

The map length of an interval of the chromosome is defined as the expected number of crossovers in this interval for a single chromatid. The probability that two alleles at the two ends of the interval come from different parental chromosome is called the recombination fraction. The relationship between the map length and the recombination fraction can be described by a mathematical map function. Different map functions are used to describe this relationship. Among these functions, the Morgan map function is the simplest one. Let the probability of no chiasma between two loci be p_0. The recombination fraction is given by: $\theta = (1 - p_0)/2$. Assume that there can only be one crossover in the chromosome segments. The probability of a chiasma in a segment of m map units is $2m$, and we have $\theta = (1 - p_0) / 2 = [1 - (1 - 2m)]/2 = m$ for $0 \leq m \leq 1/2$.

Another simple map function is the Haldane function. Haldane (1919) is the first scientist to consider the effects of two or more exchanges within a given genetic interval (Hawley and Walker, 2003). The Haldane function assumes that the crossovers are not dependent on each other and thus their occurrences will follow the Poisson process. The probability of no chiasma in the interval of m map units, p_0, is given by e^{-2m}. Applying the Mather's formula, we can have the Haldane map function: $\theta = (1 - p_0)/2 = (1 - e^{-2m})/2$. The inverse function is as follow: $m = -1/2 \ln(1 - 2\theta)$. These mathematical map functions can be computed easily with computer programs, like MAPFUN, which allow the conversion between map distance and recombination fraction.

5.1.1.2 Association Studies

The association studies work with the comparisons of the allele frequency in a disease population with that in a control population. Any significant differences between these two populations may indicate that the locus under consideration is potentially related with the disease phenotype. This association can be direct and indirect. In case of direct association, the polymorphism may have some functions that can cause/affect the disease. On the other hand, the polymorphism may not be functional for the disease but is close enough to a polymorphism causing the disease. This co-occurring of alleles more frequently than by chance is named linkage disequilibrium (LD), which is also often called "allelic association". The existence of the linkage disequilibrium can be useful for mapping disease genes. This is because these two may be very close to each other. Thus, the LD analysis has the prospect of identifying and then narrowing candidate regions of medical importance.

Compared with the linkage analysis, the association studies have several advantages. Firstly, the data can be obtained more easily than the extended pedigrees. Second, the power of detection for small genetic effects is greater. Spielman et al. (1993) showed that the insulin gene had strong association with type 2 diabetes, but with only very weak linkage. Collins et al (2004) also pointed out that linkage maps performed dissatisfactory for the common diseases because of its poor reproducibility and low power. Lastly, the range of LD is usually tens of kilobases only and can allow intensive studies on a much finer scale. This availability of the data within much smaller interval enables us to search the genome in greater details.

5.1.1.3 Markers for Linkage Analysis and Association Studies

A genetic marker is said to be useful when it can be scanned easily and reliably in the laboratory and also be highly variable. The requirement of high variability is to ensure that unrelated individuals are likely to have different alleles. The classical markers are those that are employed in the analysis of Mendelian traits. When the trait is Mendelian, we can infer the underlying genotype form the phenotypic trait, and the phenotypic trait can act as an indicator of the underlying genotype. Examples of these classical markers include ABO blood groups, and colour blindness. Other markers include the RELP markers, and Hypervariable markers etc.

The marker simple tandem repeat (STR) has been popular for the linkage analysis because its level of heterozygosity is high with increased information. SNPs have become a popular choice for association studies. It is because they are very abundant and of binary nature. Kruglyak and Nickerson (2001) estimated that there existed 7 million SNPs in the human genome with minor allele frequency (MAF) greater than 5%. Its binary nature can enable the genotyping process to become automated and high-throughput. Large-scale SNP discovery projects such as the SNP consortium (Altshuler et al., 2000) and the Hapmap project have been carrying out. The number of known SNPs has increased rapidly and such kind of data can be obtained from dbSNP etc. The dbSNP is a popular public polymorphism database.

For the studies of the genetic variations, different approaches of employing different markers have been developed, including Restriction Fragment Length Polymorphisms (RFLPs) (De Martinville et al., 1982), Amplified Fragment Length Polymorphisms (AFLP) (Vos et al., 1995), Microsatellites (or Short Tandem Repeat Sequences

(STRs)) (Taylor et al., 1989), and Single-Nucleotide Polymorphisms. Among these approaches, it is generally accepted that SNPs are of the most importance because of its abundance, stability, and simplicity (Kwok, 2002). Furthermore, it is relatively easy for designing assay for SNPs and for the subsequent scoring of these markers. As a result, nowadays, in almost every field of the genomic analysis, there are applications with the SNPs, as the SNPs are very suitable for large-scale studies that require high accuracies and high-throughput outputs. These studies include pharmacogenomics, linkage studies, and candidate gene association studies.

5.1.1.4 Their Limitations

The linkage analysis has its limitations. When combining data from different families, it will result in reduction of the genetic interval under study. More importantly, complex genetic diseases are usually caused by the combined effect of multiple polymorphisms in a number of genes and the identification of these genes by the linkage analysis has been largely unsuccessful. It is because each gene makes a small contribution to disease susceptibility and these effects likely fall below the detection threshold by the analysis unless a huge sample sizes were available (Risch, 2000).

There are also difficulties for the association studies. In the association studies, a much more number of markers are required. It is guessed that probably 30,000 to 300,000 markers (Collins et al., 1999) are required for an association study of the genome. The cost of scanning such a large amount of markers was high enough to limit the applications of the association studies in the past, while, it is expected that, with the advance of technologies today, the generation of high-density SNP maps can be efficiently realized (Antonellis et al., 2002). Association studies of this large scale have been becoming possible.

5.1.2 Constructing Linkage Disequilibrium Maps (LD Maps) with the Parametric Approach

5.1.2.1 Motivations for Constructing LD Maps

As said, the linkage maps can provide us with the information of how far two loci are from each other, in term of the number of recombination. Since the 1980s, these maps have played a key role for the positional cloning of many major disease genes. Nevertheless, when employing the linkage maps for common disease genes, where the individual phenotypic effect is smaller, the linkage methods are found not to perform well, and to be plagued by its low power. In face of these difficulties of linkage maps, their attention has focus on the exploitation of the allelic association (LD mapping). This is because the recent studies, like the one by Risch and Merikangas (1996), have shown that the power for detecting disease determinants of relatively small effect is larger with the allelic association, over the linkage.

Because of the above reasons, the linkage disequilibrium (LD) maps are developed with the aim that it can have similar features of the linkage maps, while at the same time with the capability that it can provide us with a much finer picture of the genome. It is because, as said previously, the SNPs are of much higher density than the traditional markers used in the genetic maps. The studies on the LD maps (Collins et al., 2004; Maniatis et al., 2005) suggest that, in both of these two different maps, we can have the similar abundant and narrow recombination hot spots. The LD maps can be used for the medical studies like positional cloning and evolution studies.

5.1.2.2 Theoretical Background for the Parametric LD Maps

The first LD map is proposed by Maniatis and his colleagues (Maniatis et al., 2002). It is the only common LD map up to date in our literature search. Maniatis et al were encouraged by the success of the genetic linkage maps. Maniatis et al developed the LD maps, basing on the population genetics theory for determining the expected association between the loci. Let the expected association between two diallelic loci be expressed by the probability p_t, where $0 \le p_t \le 1$, and t denotes the number of generation after founders. The association probability at generation t is:

$$p_t = (1-v)(1-\theta)\left[\frac{1}{2N_{t-1}} + (1 - 1/2N_{t-1})p_{t-1}\right]$$

where $1/2N_{t-1}$ is the probability that a random haplotype in generation t is identical by descent from a specified haplotype in the $t-1$ generation. N_t refers to the effective population size in the t generation, v is linear pressure toward linkage equilibrium from migration and mutation, and θ is the recombination frequency. This recurrence relationship can be shown to satisfy:

$$p_t - L = (p_0 - L)(1-v)^t(1-\theta)^t \prod_{i=0}^{t-1}(1 - 1/2N_i)$$

where L is the association as the number of generations approach infinity. The variable N here is for describing the stochastic variation, and v is used to make the allele frequencies keep realistically close to their present values.

Replacing $\prod_{i=0}^{t-1}\left(1 - 1/2N_i\right)$ with $e^{-t/2N}$, where N is the unknown effective size over the unknown sequence $N_0, N_1, \ldots, N_{t-1}$, and taking $(1-\nu)^t(1-\theta)^t$ as $e^{-(\nu+\theta)t}$, as t is large enough compared with v and θ, we can have:

$$p_t = p_{rt} + p_{ct}$$

where

$$p_{rt} = p_0 e^{-\left(1/2N + \nu + \theta\right)t}$$

and

$$p_{ct} = L\left[1 - e^{-\left(\frac{1}{2N} + v + \theta\right)t}\right]$$

When N is a constant, we can have p_t expressed as:

$$p_t = (1-L)Me^{-\theta t} + L$$

where M is defined as:

$$M = (p_0 - L)e^{-(v+\frac{1}{2N})t} / (1-L)$$

The Malecot equation is obtained by the assumption that the role of the above θt can be replaced with εd, where ε is assumed to be constant for a specified region, and d is the distance between loci, which can be measured in genetic scale or physical scale. The association at distance d can be expressed as:

$$p_d = (1-L)Me^{-\varepsilon d} + L$$

It should be noted that, in the Malecot model, there are four parameters that are required to be estimated. They are M, L, ε and S, the location of a disease locus as function of distance. This model can be found unsuitable when there are any significant evolutionary variance, non-independence of samples, variable recombination, map error, or any other departure from the model's assumption.

5.1.2.3 Constructing the Parametric LD Maps

A map interval can be defined with a pair of DNA sites, which may sometimes be called "markers". The association probability between two marker alleles can be estimated from the theory of estimating the covariance D for a random sample of haplotypes:

$$p = \frac{D}{Q(1-R)}$$

where Q denotes the frequency of the rarest (youngest) allele, R is the frequency of the associate marker allele, and D refers to the absolute value of the difference between a haplotype frequency and its equilibrium value. The equilibrium values can be obtained as the product of allele frequencies.

The Malecot model can also provide us with the estimation of the association probability:

$$p_d = (1-L)Me^{-\varepsilon d} + L$$

Then, a natural measure for LD is the εd. Collins et al. (2004) pointed out that εd is not biased and can be much more accurately known than the θt in the previous population model. This parameter ε is non-negative and depends on the number of generations that the haplotypes need for approaching equilibrium. Its inverse, 1/ε, is termed by Collins et al. as the 'swept radius', which is the distance in kilobases that make M reduced by the factor e^{-1}. The swept radius defines the extent of 'useful' LD. Useful LD means the extent where LD is useful for gene mapping.

Maniatis et al. found that the parameter L can only be estimated inaccurately when computing over small distances, together with the estimation of the ε. It would become useful to decide an independent estimate of L, which is the mean value of p as $e^{-\varepsilon d}$ approaches zero. The swept radius of the model can be estimated by fitting multiple pair-wise measures of association probability p into the Malecot model with composite likelihood. The mean swept radii are estimated to be in the range of 30–56 kb for Caucasians and Asians. That for the African-Americans is 22–41 kb.

After obtaining the LD map lengths between each pairs of consecutive SNPs, a map can be simply constructed for these m SNPs. Let the length for the ith interval be $\varepsilon_i d_i$ LD units (LDU). The unit LDU is defined here to be equal to one swept radius. Then, the total LD length for the interval between the first and the last SNP is given by $\Sigma\varepsilon_i d_i$ LDU, and that for the physical distance is Σd_i kb. The ratio of these two distances can be used as a rough estimate of the ε in that region.

Collins et al. shown that the locations of the regions of intense recombination correspond closely to the steeper segments ('steps') on the LD map, whereas the cool recombination areas appear as flat segments ('plateaus') on the LD map. When the SNPs are separated with large distances, there is no useful LD and it can be said that these pairs are mostly uninformative. Thus, it can be viewed that there is a sliding window, within where the pairwise association data are informative for the particular map interval.

5.1.2.4 Advantages and Applications of the LD Maps

In the LD maps, different regions of medical interests can be shown conveniently and visually. A cold spot for LD is an interval in the LD map where LD declines rapidly with distance. When the LD maps are used to mirror the recombination, the cold spot for LD is a hot spot for recombination and vice versa. The hot spot for LD is the region in the LD map that appears as a flat region, which LD declines slowly against distance. It should be pointed out that the corresponding relationship between the LD and the recombination is not always so well defined, as there are other unpredictable genetic factors that may influence this corresponding.

Besides showing the recombination patterns in the LD map, the LD map can facilitate the optimal spacing of the tag SNPs. In order to have enough coverage of a genomic segment, it is required that there will be tag SNPs within each LD unit. In the study of the MHC class II region in UK, of Sammi and Zimbabwean populations, Kauppi et al. (2003) found that all the populations showed very similar LD patterns. This result was also supported by other studies of Lonjou et al. (2003) and Gabriel et al. (2002). Thus, Collins et al. suggested that a 'standard' LD map of the populations is possible. Another advantage of the LD map is its increase of the power for disease gene localization. Collins et al. estimated in the simulation study that the mean power of using the kilobase map is only about 62% of the power of using the LDU map for the disease gene localization.

5.2 Theoretical Background for Non-parametric LD Maps

5.2.1 Formulating the Non-parametric LD Maps Problem as an Optimization Problem with Quadratic Objective Function

When two diallelic SNPs A and B are said to be linkage equilibrium, it means that their distributions are the same as Table 5.1 shown below:

where the p is the probability of obtaining an allele A from the pool of alleles A and a. Similarly, q is that of getting an allele B from the pool of alleles B and b. Usually, the probability p is used to denote the allele with the lowest probability and q for the second lowest one, so we have the following relationship: $p \le q, p \le 0.5, q \le 0.5, p \le 1 - q$.

When the two SNPs are in linkage disequilibrium (LD), it can be represented by the parameter d, which is defined as followed (Sham, 1998):

Here, d is defined in such a way that it remains non-negative all the time. When d is equal to zero, then the above table (Table 5.2) is reduced to that for the linkage equilibrium case.

The unit for measuring the scaled linkage disequilibrium is *d'* and is defined by:

$$d' = \frac{d}{p(1-q)}$$

Table 5.1 Distribution table for the two alleles A and B in linkage equilibrium

	Allele B	Allele b	Row sum
Allele A	Pq	p(1 – q)	p
Allele a	(1 – p)q	(1 – p)(1 - q)	(1 – p)
Column sum	Q	(1 – q)	1

Table 5.2 Distribution table for the two alleles A and B in linkage disequilibrium

	Allele B	Allele b	Row sum
Allele A	pq + d	p(1 – q) – d	p
Allele a	(1 – p)q – d	(1 – p)(1 – q) + d	(1 – p)
Column sum	Q	(1 – q)	1

As an example, let $d = 0.1$, $p = 0.2$ and $q = 0.3$, then $d' = {}^{0.1}\!/_{0.2(1-0.3)} = 0.7143$. From the theoretical population biology, it is known that the above d' decays by a factor of $1 - \theta$ per generation, where θ is the recombination fraction. Thus, we can have: $-\ln d'^2 \propto -\ln(1-\theta)$. For small values of θ, it is approximate true that $-\ln d'^2 \propto \theta$, and, therefore, $-\ln d'^2$ is approximately close to the genetic map distance measured in units of Morgan (Sham, 1998).

Because of the above property of the $-\ln d'^2$, it is used to measure the LD distance between each pair of SNPs. Consider the case of n SNPs, we can have the n-by-n distance matrix for representing the distances between each pair of these n SNPs, with the (i, j) element defined as $[-\ln d_{ij}'^2]$. It is noted that this distance matrix is symmetric and has the diagonal elements equal to zeros. For applications in the LD maps, it is desirable to have the representation of these distances in one dimension. Also, the biological knowledge tells us that it is desirable to have the preservation of original ordering of the SNPs in the chromosome. This is because physically closer SNPs have smaller genetic map distances between them too. The objective of this scaling into one dimension is that this scaled distance values should be in agreement with the original distances in the n-by-n distance matrix as much as possible. This objective is very similar with the classical unidimensional scaling problem.

In the classical unidimensional scaling problem, it is to place n objects on the real line, such that the interpoint distances can best approximate the observed dissimilarities between pairs of objects. It is well known that this problem is equivalent to an NP-hard combinatorial problem. Our constrained unidimensional scaling problem here is to place *n* objects in a given order on the real line, such that the interpoint distances best approximate the observed dissimilarities between pairs of objects. In the literature, researchers are interested in constrained multidimensional scaling problems. Lee (1984) used least squares scaling to allow not only for equality constraints but also inequality constraints.

5.2.2 Mathematical Formulation of the Objective Function for the LD Maps

Mathematically, the above constrained unidimensional scaling problem can be expressed with the objective function J defined as:

$$J(z_1, z_2, \ldots, z_{n-1}) = \sum_{i=1}^{n} \sum_{j=1}^{i-1} w_{ij} \left(d_{ij} - \sum_{k=j}^{i-1} z_k \right)^2$$

where d_{ij} denotes the original distance between the *i*th SNP and *j*th SNP in our distance matrix, and z_k denotes the scaled distance between the *k*th SNP and the $(k+1)$th SNP. The weighting coefficients w_{ij} can be defined as, for example, the length of the 95% confidence interval of the distance $-\ln d_{ij}'^2$. Mathematically,

$$w_{ij} = \frac{1}{-\ln CIL_{ij}^2 + \ln CIH_{ij}^2}$$

where ($CIL_{ij} > 0.05$ or $CIL_{ij} < d_{ij}'$) and ($CIH_{ij} > CIL_{ij}$); otherwise, $w_{ij} = 0$. CIL_{ij} represents the lower 95% confidence interval, and CIH_{ij} represents the upper 95% confidence interval of the estimated d_{ij}. When the 95% confidence interval is narrower, it means that we know quite sure about the distance between the SNP pair. Thus, their weighing should be larger. It is also noted that w_{ij} is non-negative.

We also have the following constraints:

$$z_k \geq 0, k = 1,2,\ldots,n-1$$

This is to ensure that the scaled distances are in consistence with the orderings of SNPs in the chromosome. The scaled distance between two non-consecutive SNPs is defined as the summation of the distances between all the consecutive SNP pairs that can be found between these two SNPs. Mathematically, $z_{j(i-1)} = \sum_{k=j}^{i-1} z_k$, where $i > j$, and the new symbol $z_{j(i-1)}$ is for representation of the scaled distance between SNP j and SNP (i - 1).

5.2.3 Constrained Unidimensional Scaling with Quadratic Programming Model for LD Maps

It can be recognized that, with the above formulation of the objective function J, the above problem is the one with the quadratic optimization and the optimal solution of J can be obtained through quadratic programming techniques. Quadratic programming refers to the finding of the optimal value of the problem with a quadratic objective function and linear constraints. Problems of quadratic programming are important in their own right (Nocedal and Wright, 1999). A formulation for the general quadratic program (QP) is as followed:

$$\min_x q(x) = \frac{1}{2} x^T G x + x^T d$$

subject to $a_i^T x = b_i$, $i \in$ E, and $a_i^T x \geq b_i$, $i \in$ I. G is a symmetric n-by-n matrix, E and I are finite sets of indices. $\{a_i\}$, where $i \in E \cup I$, d and x are vectors of n elements.

The active-set methods are found to be effective for small- to median-scale QP problems (Nocedal and Wright, 1999). The method will start with making a guess of the optimal active set. If the guess is found incorrect during iteration, gradient and Lagrange multiplier information will be used to guide the dropping of one index from the current estimate and then a new index will be added. They are different from the simplex method, which requires the iterates to move from one vertex of the feasible region to another. In active-set methods, some iterates may lies at the interior of the feasible region. Primal, dual and primal-dual methods are the three active-set methods for solving QP. A disadvantage of these methods is that, in each iteration, there is usually only a single change and thus it may require many iteration steps to converge on large-scale problems.

The gradient projection is designed with the mind that it can overcome the slow convergence problem of active set methods. And the gradient projection makes rapid changes to the active set. In each iteration, the steepest descent direction from the current point is searched. If there is a bound, the search direction is corrected so that it stays feasible. A local minimizer along this direction will then be located and is called Cauchy point. The working set is now the set of bound constraints that are active at the Cauchy point. Then, the sub-problem that has the active components fixed will be solved to "explore" the face of the feasible box on which the Cauchy point lies.

Interior-point methods have the name because they require that, in each iteration, the inequality constraints of the optimization problem have to be satisfied strictly. They can be applied for solving the QP problems. It has the advantages of simple description, relatively easy implementation and quite efficiency.

It can be observed that our quadratic objective function has the following property:

Property 5.1 The objective function value J is equal to zero if and only if

$$d_{ij} = \sum_{k=i-1}^{j} d_{k(k+1)}$$

for all $i < j$.

For our above constrained optimization problem, it can be re-written as the following least squares problem with non-negativity constraint:

$$\min_{z \geq 0} \| WAz - Wd \|_2^2$$

where $z = (z_1, z_2, \ldots, z_{n-1})^T$ and $d = (d_{21}, d_{31}, \ldots, d_{n1}, d_{32}, d_{42}, \ldots, d_{nn-1})^T$,

and where $W = diag(w_{21}, w_{31}, \ldots, w_{32}, w_{42}, \ldots, w_{nn-1})$, $A = \begin{pmatrix} A_1 \\ A_2 \\ \vdots \\ A_{n-1} \end{pmatrix}^T$.

A_i is an $(n-i)$-by-$(n-1)$ matrix given by $[0_i | T_i]$, 0_i is an $(n-i)$-by-$(i-1)$ zero matrix, T_i is an $(n-i)$-by-$(n-i)$ Toeplitz matrix with its first column $[1,1,\ldots,1]^T$ and its first row $[1,0,\ldots,0]$.

The MATLAB function that we employed for solving this least squares problems with non-negativity constraint is the lsqnonneg function. It uses the algorithm described by Lawson and Hanson (1974). In the algorithm, it will start with a set of possible basis vectors. Then, the associated dual vector will be computed, and let this dual vector be called lambda. The basis vector is selected with the one that corresponds to the maximum value in lambda, and that can swap the lambda out of the basis in exchange for another possible candidate. The algorithm will continue until lambda becomes less than or equal to zero. Generally speaking, this algorithm is efficient for solving the optimization problem (MATLAB, 2005).

5.3 Applications of Non-parametric LD Maps in Genomics

The above quadratic programming method finds the scaled distances in each SNP intervals in terms of linkage disequilibrium units. The scaled distances are the basis for constructing linkage disequilibrium maps. The LD map is analogous to the linkage map, nevertheless, with substantial difference by accommodating recombination events that have accumulated (Jeffreys et al., 2001). In the following sections, the differences and similarities with the linkage disequilibrium patterns between populations and chromosome regions are illuminated with our proposed method. The computational properties of our proposed method are also investigated.

5.3.1 Computational Complexity Study

The MATLAB quadratic programming solver is employed for obtaining the optimal solution of this constrained unidimensional scaling problem. The SNP data is the ENCODE dataset from the HapMap project (30 October 2004) and is based on NCBI build34. The entries of the original distance matrix are obtained with running the program HAPLOVIEW on these ENCODE dataset and with a simple C script of transforming the d'_{ij} values into $-\ln d'^{2}_{ij}$ values.

The results of Tables 5.3 and 5.4 are run on MATLAB program on the Linux platform of CPU Intel 3.2c with 1G memory. From the above tables, it is observed that the computational time for the above quadratic programming algorithm is of $O(n^4)$, and that the memory requirement for the inputs are of $O(n^3)$. The following figures (Figs. 5.1–5.4) show the scaled SNP positions against the original physical

Table 5.3 Computational time (seconds) for segments of chromosome 9q34 of different length with quadratic programming algorithm

TimeSNP No.	54	107	213	426
Time	0.18	2.15	85.87	1,747

Table 5.4 Computational time (seconds) for chromosomes where quadratic programming algorithm successfully

	chr7p15	chr8q24	chr9q34	chr18q12
SNPs No.	466	533	426	536
Q.P. Time	2,720	8,993	1,747	12,030

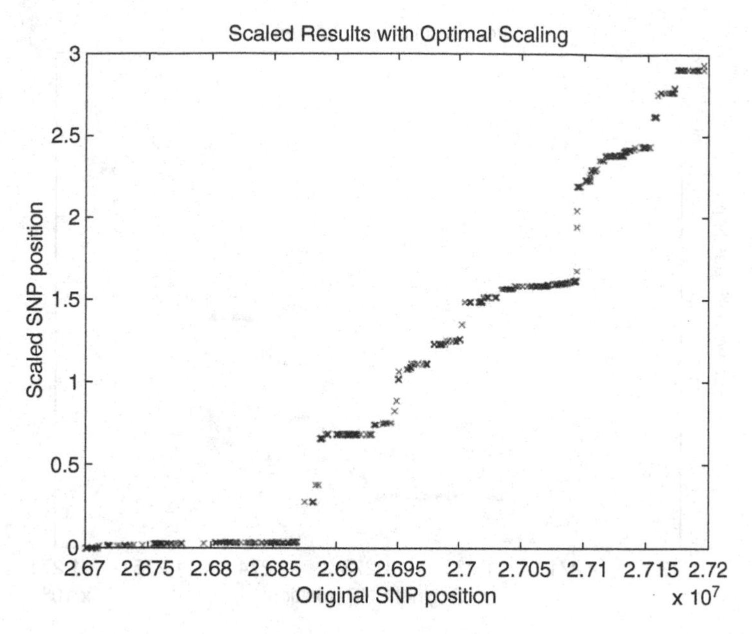

Fig. 5.1 Scaled SNP position vs. original SNP position for chr7p15 with the quadratic programming algorithm

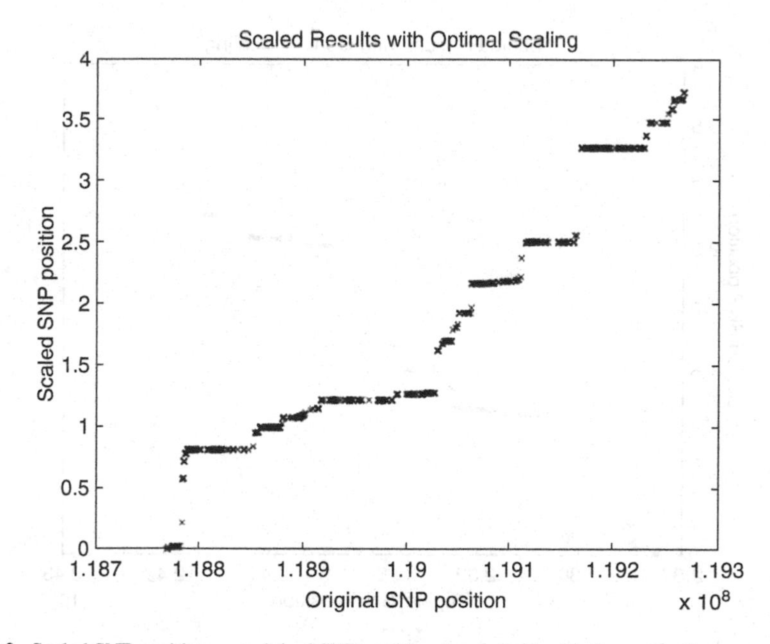

Fig. 5.2 Scaled SNP position vs. original SNP position for chr8q24 with the quadratic programming algorithm

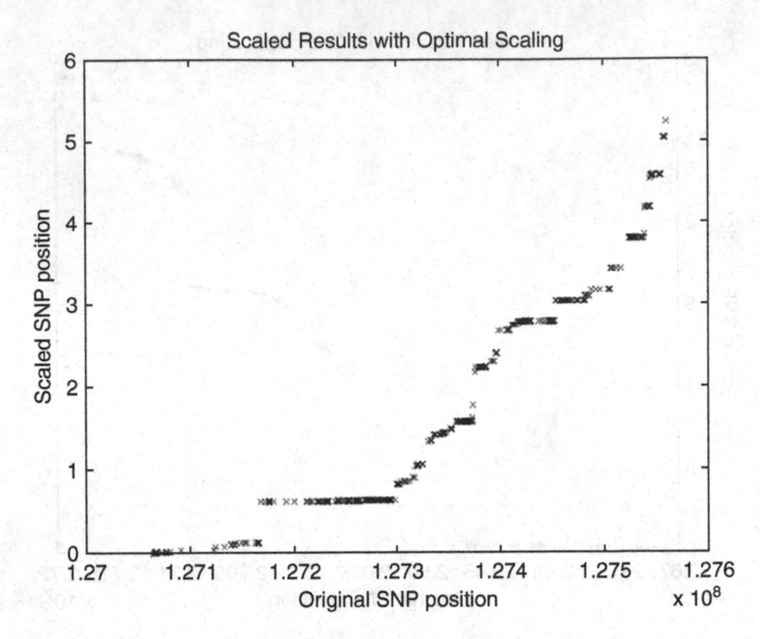

Fig. 5.3 Scaled SNP position vs. original SNP position for chr9q34 with the quadratic programming algorithm

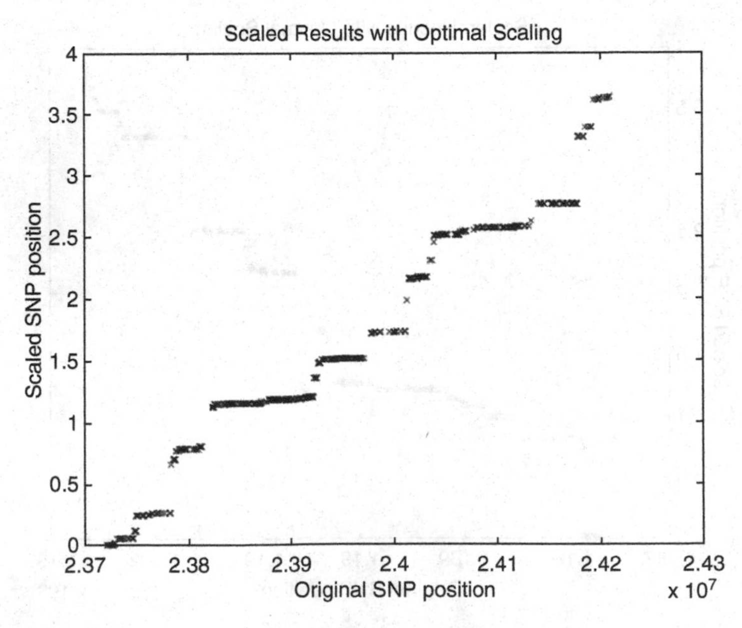

Fig. 5.4 Scaled SNP position vs. original SNP position for chr18q12 with the quadratic programming algorithm

positions. From these graphs, we can be easily identified the hot recombination regions and cold recombination regions.

For other larger chromosome regions, like chromosome 12q12, which is the fifth smallest region in term of the number of SNPs and have 665 SNPs, the quadratic programming approach can not be run successfully with problems of the out of memory errors.

5.3.2 *Genomic Results of LD Maps with Quadratic Programming Algorithm*

The maps below are constructed with our proposed method for the different human genome regions. From the maps, it can be helpful for identifying informative and uninformative pairs of SNPs. For SNPs that are separated by large scaled distances, it means that there exist no useful LD and, as a result, these pairs are uninformative. The steeper segments on the linkage disequilibrium map correspond closely to the location of the recombination intense regions. On the other hand, the high fairly flat lands in the LD map correspond to the recombination cool areas.

To observe the linkage disequilibrium of the different populations, we also applied our algorithm to all the four different populations in the chromosome 9 of the Hapmap ENCODE dataset as an example. The following figures (Figs. 5.5–5.8) give us the scaled LD maps of these four different population samples, i.e., the

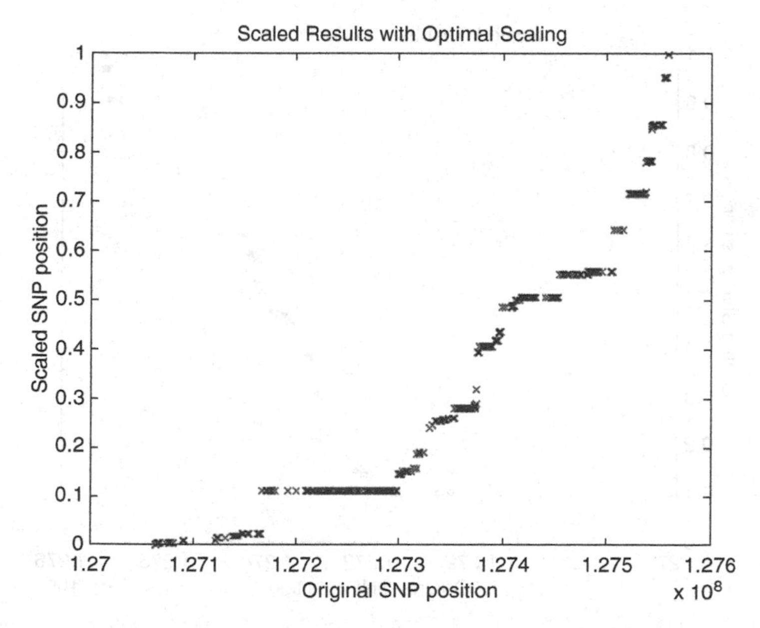

Fig. 5.5 The scaled SNP position via the original SNP position, CEU samples

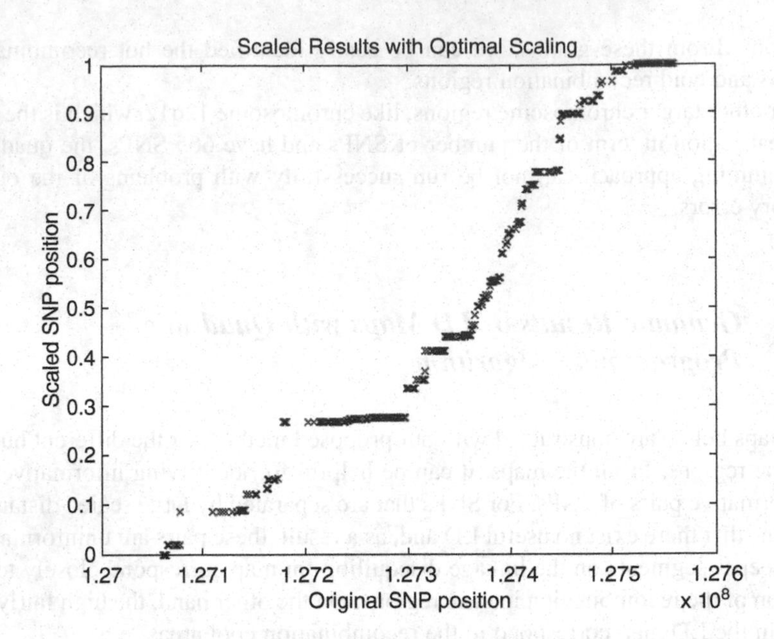

Fig. 5.6 The scaled SNP position via the original SNP position, YRI samples

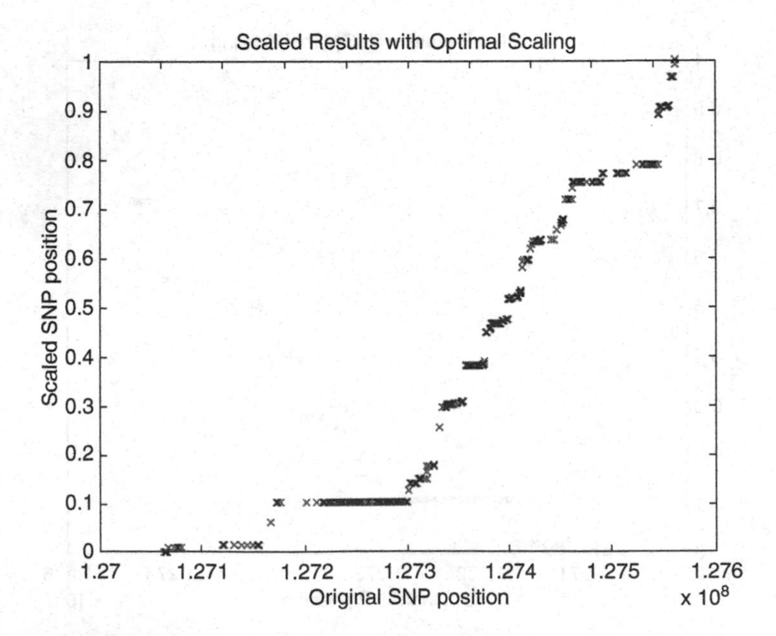

Fig. 5.7 The scaled SNP position via the original SNP position, HCB samples

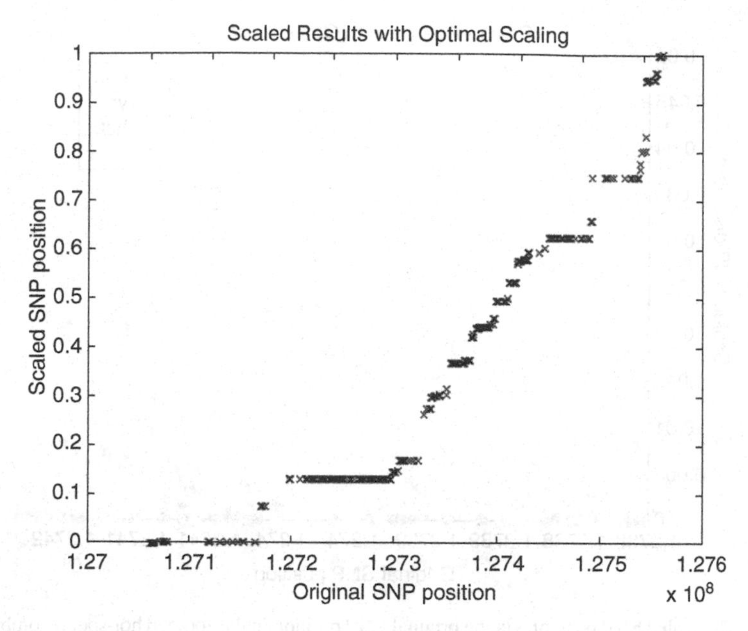

Fig. 5.8 The scaled SNP position via the original SNP position, JPT samples

scaled SNP position via the original SNP position. The four populations are CEU of 444 SNPs, YRI of 476 SNPs, HCB of 399 SNPs and JPT of 391 SNPs in the region of the chromosome 9 from the HapMap Project. The CEU population refers to the samples that were collected in U.S. residents with northern and western European ancestry by CEPH. The YRI refers to the samples of the Yoruba people of Ibadan, Nigeria. The HCB and JPT are the samples of Beijing, China and Tokyo, Japan. In the figures, the flat regions are called the cold-spot recombination regions, which mean that a recombination in the meiosis process seldom occurs. The steep regions in the figures are the hot-spot recombination regions, which mean that a recombination in the meiosis process occurs frequently.

In these four populations (CEU, YRI, HCB and JPT), 86.23, 85.26, 84.92 and 87.69% of the total regions are cold-spot recombination regions respectively (defined here as the regions of zero LD distance with its neighboring SNPs). According to the above figures, we can observe that the positions of the cold-spot recombination regions in the four scaled LD maps from the four populations are about the same. Furthermore, we identify several hot-spot recombination regions in the chromosome interval between positions 127380000 and 127420000 for the four populations. We define a hot-spot recombination region to be a region where has nonzero LD distances with its neighborhood SNPs. The figure below (Fig. 5.9) depicts the zoomed scaled LD distances of the SNPs in these hot-spot recombination regions. Even though all of the four populations have about the same hot-spot recombination regions, their corresponding nonzero scaled LD distances appear in different chromosome positions. It is interesting to further investigate the relationship between these LD patterns and population demography.

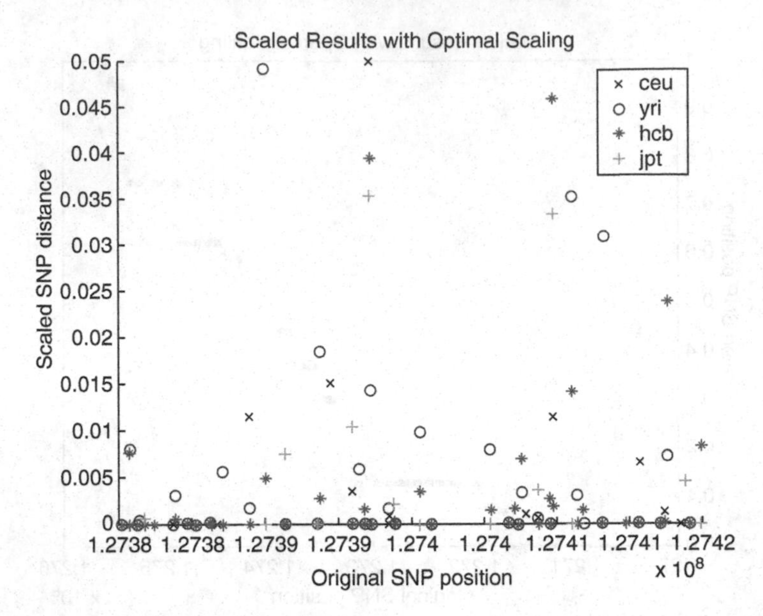

Fig. 5.9 The scaled SNP position via the original SNP position in the zoomed hot-spot recombination region of chromosome interval between 127380000 and 127420000 for four population samples

For comparison, the following figures (Fig. 5.10–5.13) give us the scaled SNP position via the cumulated LD distances with respect to the above Figs. 5.5–5.8 respectively. We consider that all the sums of the cumulated LD distances are equal to 1 for the convenience of comparison. According to these figures, we see that the LD patterns of different populations have different scaled SNP position (i.e., scaled distances). For example, we observe that there is a big gap in the cumulated LD distance in JPT samples, but the corresponding scaled distances change is very small. Such plots of our scaled LD maps can give us more detailed comparisons of LD patterns with the scaled distances among different populations.

The following table (Table 5.5) lists the percentages of the overlapping of the cold-spot recombination regions between the different populations. It is observed that the pair of the HCB and the JPT populations has the highest percentage of overlapping of their cold-spot recombination regions (87.11%), whereas the pair of the CEU and the YRI populations has the lowest percentage (82.43%). The second table (Table 5.6) below lists the percentages of the overlapping of the hot-spot recombination regions between the different populations. It can be observed that the pair of the HCB and the JPT populations has again the highest percentage of overlapping of their hot recombination regions (43.05%), while the pair of the YRI and the JPT populations has the lowest percentage (18.75%). We remark that it is commonly believed that the genomes of the HCB and the JPT are very similar, while the genome of the YRI is quite different from those of the other populations.

It can also be observed that the overlapping of the hot-spot recombination regions is much less than that in the cold-spot recombination regions. This result is

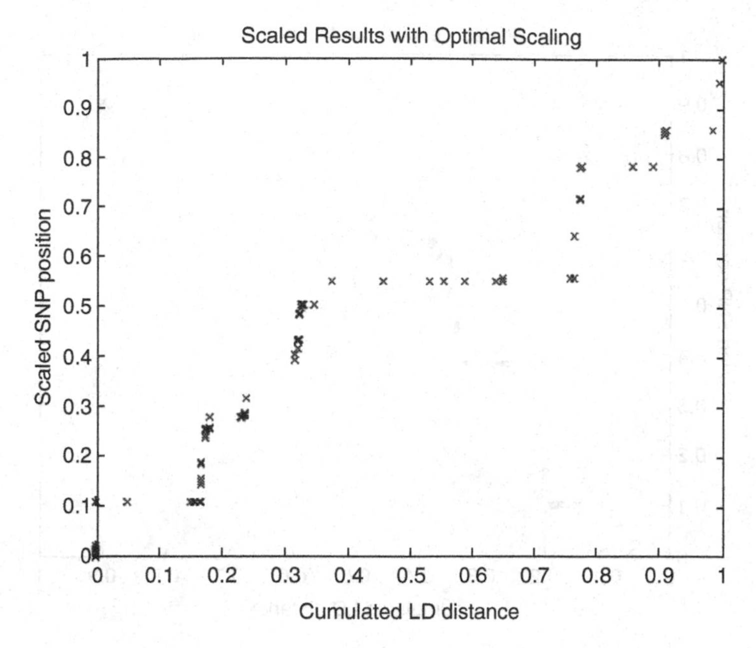

Fig. 5.10 The scaled SNP position via the cumulated LD distances, CEU samples

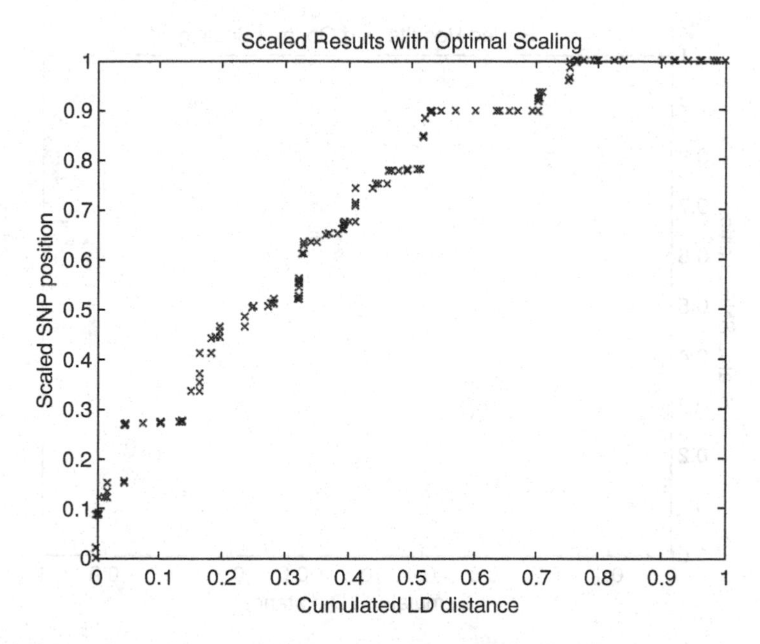

Fig. 5.11 The scaled SNP position via the cumulated LD distances, YRI samples

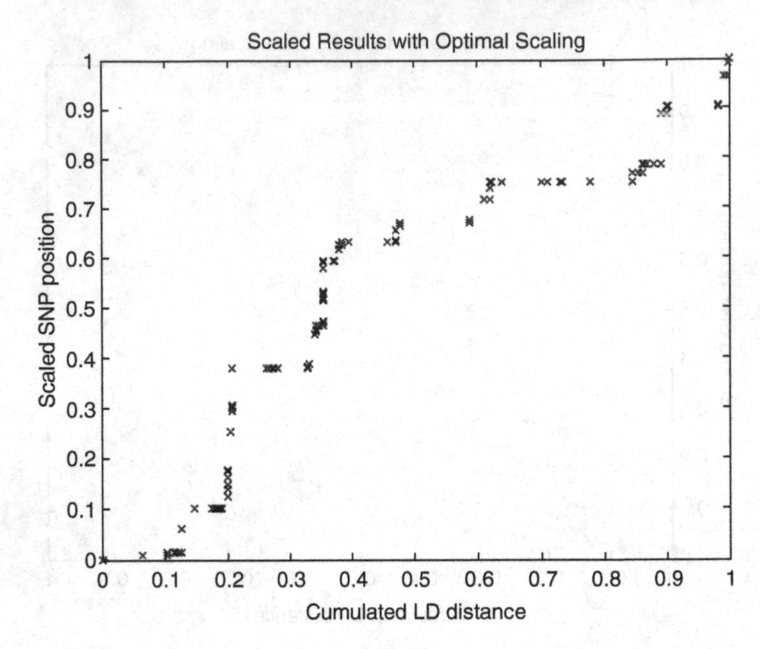

Fig. 5.12 The scaled SNP position via the cumulated LD distances, HCB samples

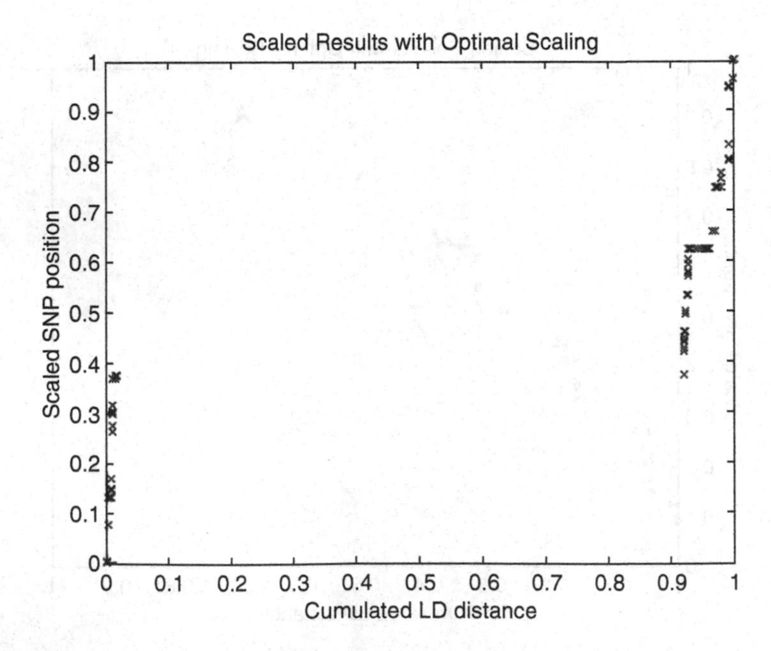

Fig. 5.13 The scaled SNP position via the cumulated LD distances, JPT samples

Table 5.5 Overlapping of the cold-spot recombination regions in the four populations

Populations	Overlapping of the cold-spot recombination regions (in percentage)
CEU and YRI	82.43%
CEU and HCB	83.25%
CEU and JPT	84.60%
YRI and HCB	84.91%
YRI and JPT	84.05%
HCB and JPT	87.11%

Table 5.6 Overlapping of the hot-spot recombination regions in the four populations

Populations	Overlapping of the hot-spot recombination regions (in percentage)
CEU and YRI	17.52%
CEU and HCB	31.37%
CEU and JPT	27.74%
YRI and HCB	33.33%
YRI and JPT	18.75%
HCB and JPT	43.06%

consistent with that in the Fig. 5.9. It suggests that the hot-spot locations are different among the population samples.

For comparison, the two figures (Figs. 5.14 and 5.15) below give the scaled LD maps with the parametric approach and with our non-parametric approach. The figures refer to the single nucleotide polymorphisms mapping from an 880 kb region flanking *CYP2D6* (Hosking et al., 2002). A total of 1,018 Caucasians were genotyped for the 27 single nucleotide polymorphisms there, which were known to result in the recessive *CYP2D6* poor drug metaboliser phenotype. We can observe that the basic shapes of both figures are similar with each other. The cold-spot regions and the hot-spot regions occur in similar regions of the chromosome. Nevertheless, it can also be observed that there are some small differences in the two maps. It may be interesting to further investigate the similarity and difference between these two approaches.

5.3.3 Construction of the Confidence Intervals for the Scaled Results

Another issue that we would like to study is how the sampling process in the Hapmap dataset affects linkage disequilibrium information among SNPs. The idea is to use bootstrap method for generating several LD maps to check their shapes. In this study, we employ 100 SNPs from the human genome position 127063383 to

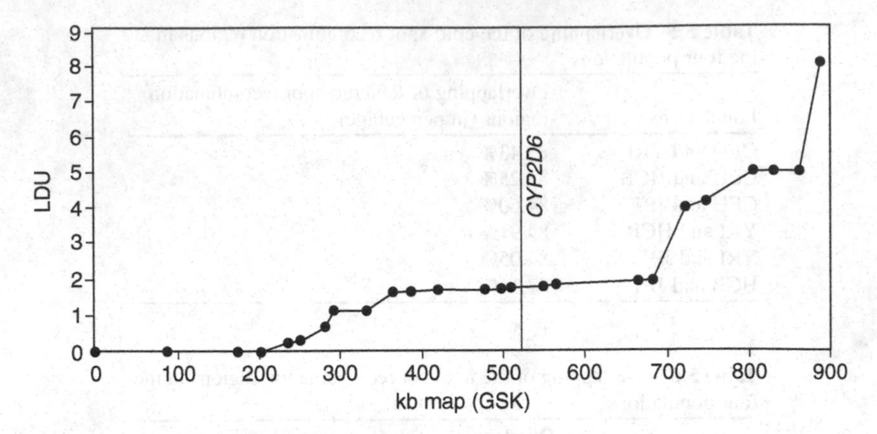

Fig. 5.14 The LD map constructed with the parametric method (Maniatis et al., 2005). Vertical line indicates the location of the locus at 525.3 kb

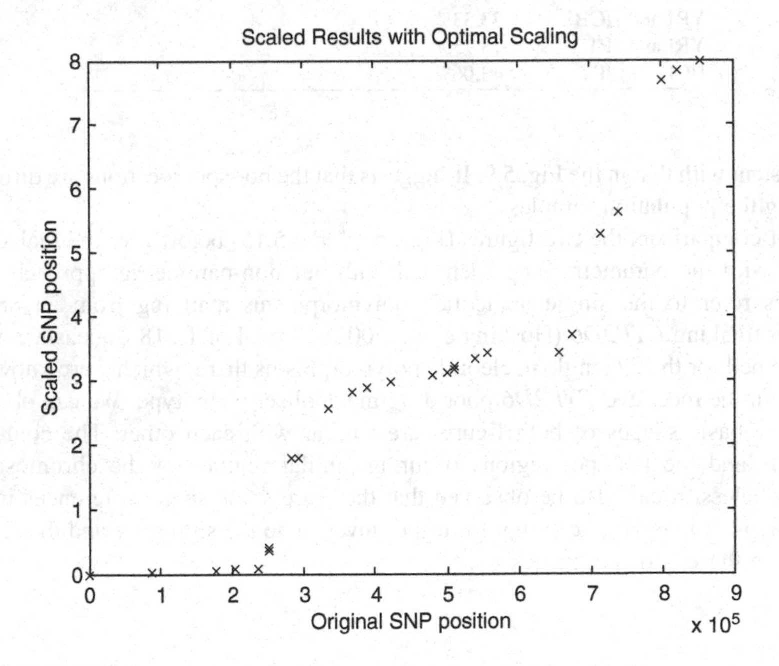

Fig. 5.15 The LD map constructed with the quadratic programming algorithm

127249267 for the demonstration. In the data set, we have 45 persons from the CHB populations. We apply the bootstrap method to generate several samples for the construction of the LD maps. A bootstrap sample is generated by random selection from the 45 persons with replacements. Since we do not want to create any new

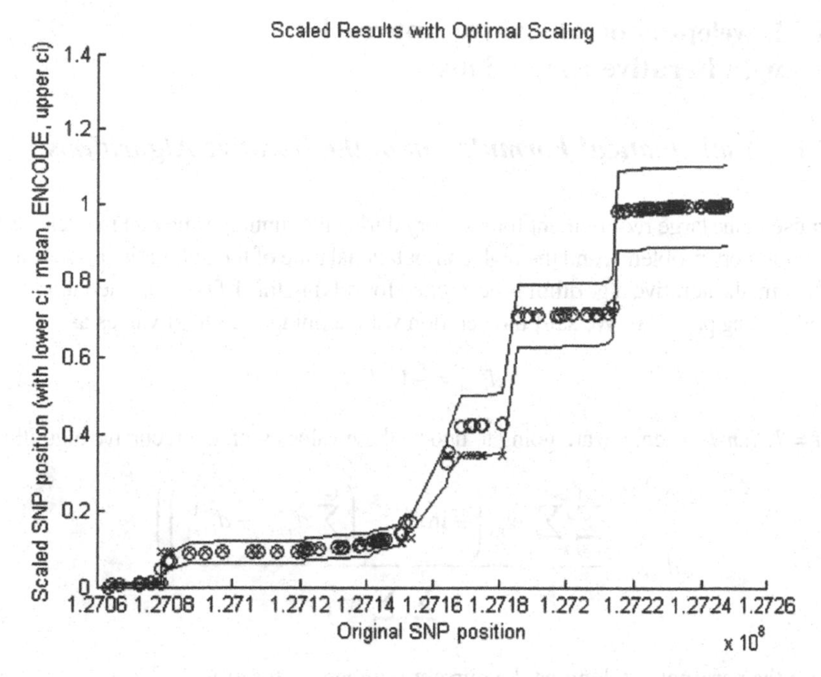

Fig. 5.16 Scaled LD position with the upper and lower confidence intervals (95%), the mean position (with 'o') and the position of the original Hapmap data set (with 'x') via the original SNP position

genetic information, we do not mix the genotypes of the 45 persons in the construction of the bootstrap samples. In our test, a total of 30 bootstrap samples are constructed. Then the 95% confidence intervals are constructed from the scaled LD maps of these bootstrap samples. The figure below (Fig. 5.16) shows the mean curve of the LD map and its confidence intervals. For illustration, the LD map by using the original Hapmap data set is also displayed.

According to this figure, some positions of the 95% confidence intervals are larger, and some are thinner. In particular, the 95% confidence intervals in the hotspot recombination regions are thinner. We can also observe that the map with the original Hapmap data set is contained in the maps of the upper and lower 95% confidence intervals. And, this LD map of the Hapmap data set is very similar with the mean LD map constructed with the bootstrap method. The average percentage difference between these two scaled LD maps is 5.92%. The only major difference between these two LD maps can be found in the small interval around the chromosome interval 127168000 to 127184000, where the LD map of the Hapmap data set is close to the bottom of the 95% confidence level. From the experimental results, it can be observed that the sampling process does not affect this chromosome region significantly, and that the LD maps can give us stable linkage disequilibrium information of the SNPs in this chromosome region.

5.4 Developing of Alterative Approach with Iterative Algorithms

5.4.1 *Mathematical Formulation of the Iterative Algorithms*

Because of the large requirement for memory during the running time that may cause the out of memory problems, and the high computational time of the quadratic programming algorithm, the iterative algorithm is developed for solving this LD constrained unidimensional scaling problem. We start the iteration with defining the initial values as

$$d_{i,i+1}^{0} = -\ln d'^{2}_{i,i+1}$$

for $i = 1,\ldots,n-1$. Then, we are going to update these values with the recursive algorithm:

$$d_{i,i+1}^{t} = \frac{\sum_{j=1}^{i-1}\sum_{k=i+1}^{n} w_{j,k}\left(-\ln d'^{\,2}_{j,k} - \left(\sum_{u=j}^{k-1} d_{u,u+1}^{t-1} - d_{i,i+1}^{t-1}\right)\right)}{\sum_{j=1}^{i}\sum_{k=i+1}^{n} w_{j,k}}$$

where the parameter *t* denotes the current number of iteration.

From the above table (Table 5.7) or from our updating algorithm, it can be observed that the computational complexity of the above iterative equation (1) is $O(n^4)$, and the memory requirement for the iterative algorithm is $O(n^2)$.

In order to simplify the computational process, the approximate iterative algorithms are developed as followed:

$$d_{i,i+1}^{t} = \frac{\sum_{j=j\_near}^{i-1}\sum_{k=i+1}^{k\_near} w_{j,k}\left(-\ln d'^{\,2}_{j,k} - \left(\sum_{u=j}^{k-1} d_{u,u+1}^{t-1} - d_{i,i+1}^{t-1}\right)\right)}{\sum_{j=j\_near}^{i}\sum_{k=i+1}^{k\_near} w_{j,k}}$$

where the variable *j_near* is the defined as max(1,*i–near*), *k_near* as min(*SNP no*, *i* + *near*). The variable *near* is the number of the near-by SNPs that are going to be used in the approximation. The complexity for the approximation algorithm is $O(n)$. We will call this approximate algorithm "iterative algorithm with *near* nearby SNPs', and the original iterative algorithm "iterative algorithm with all nearby SNPs'.

Table 5.7 Computational time (seconds) for chromosomes with the iterative algorithm, compared with the quadratic programming algorithm

	chr7p15	chr8q24	chr9q34	chr18q12
SNPs No.	466	533	426	536
Q.P. Time	2,720	8,993	1,747	12,030
All Nearby SNPs Time	5,662	9,066	3,785	9,198

5.4.2 Experimental Genomic Results of LD Map Constructions with the Iterative Algorithms

5.4.2.1 Experimental Scaled Results for the Iterative Algorithm with All Nearby SNPs

Figs. 5.17–5.20 show us the scaled SNPs position against their original position with the iterative algorithm with all nearby SNPs for the chromosome regions chr7p15, chr8q24, chr9q34 and chr18q12. We can observe that the scaled positions are almost exactly the same as those from the quadratic programming algorithm. Thus, it suggests that both algorithms can find the optimal solutions for the constrained unidimensional scaling problem.

5.4.2.2 Computational Time Results for the Iterative Algorithms

As we have discussed in the previous section, the approximate iterative algorithms have been developed because the computational time for the iterative algorithm with all nearby SNPs are still of order $O(n4)$. From the two tables below (Tables 5.8 and 5.9) and from the counting of steps of the approximate iterative algorithms, we can find that the computational time is now of $O(n)$. The memory requirement is still the same as that of the iterative algorithm with all nearby SNPs.

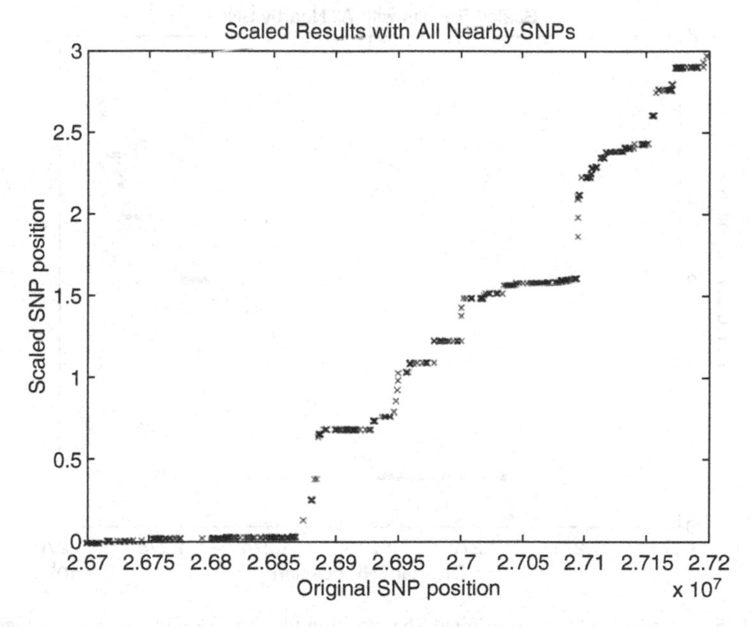

Fig. 5.17 Scaled SNP position vs. original SNP position for chr7p15 with the iterative algorithm of all nearby SNPs

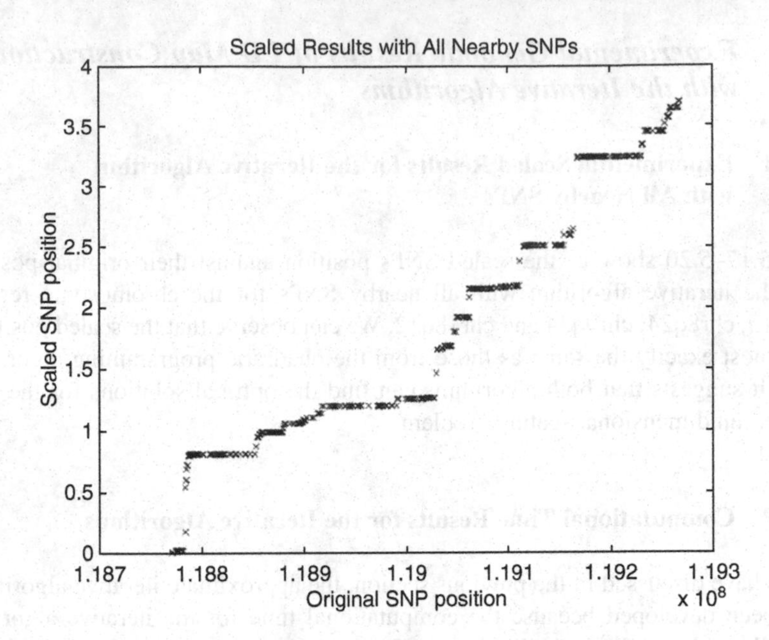

Fig. 5.18 Scaled SNP position vs. original SNP position for chr8q24 with the iterative algorithm of all nearby SNPs

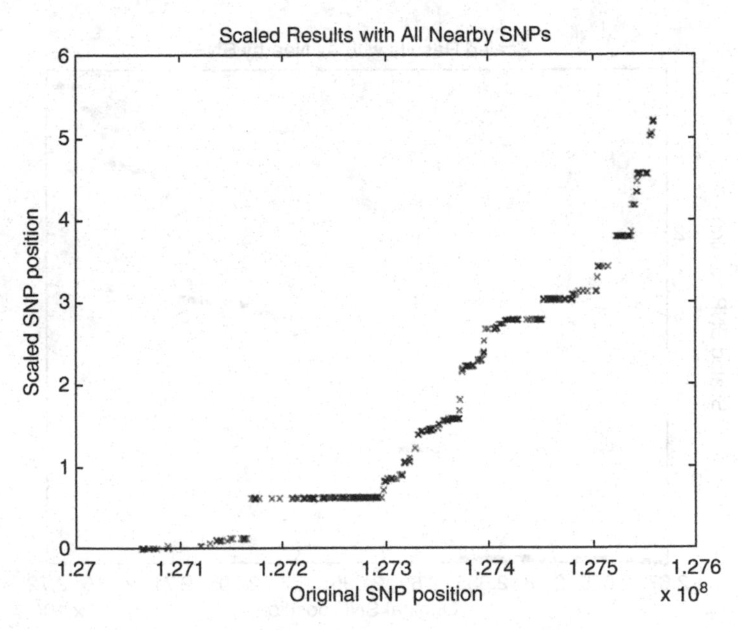

Fig. 5.19 Scaled SNP position vs. original SNP position for chr9q34 with the iterative algorithm of all nearby SNPs

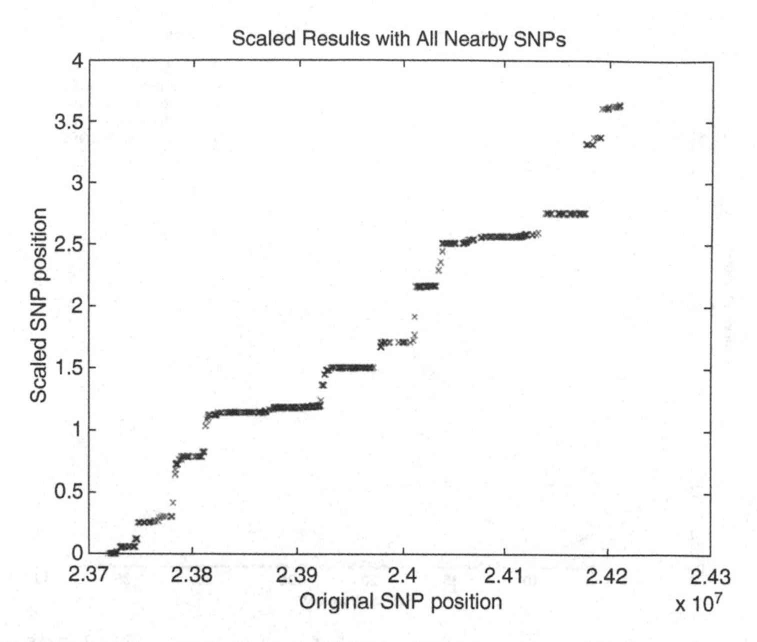

Fig. 5.20 Scaled SNP position vs. original SNP position for chr18q12 with the iterative algorithm of all nearby SNPs

Table 5.8 Computational time (seconds) for chromosome regions (with SNP number) by the iterative algorithms with the number of nearby SNPs set at different values

	5 nearby	10 nearby	20 nearby	30 nearby	All nearby
chr7p15 (466)	11.86	63.16	393	1,234	5,662
chr8q24 (533)	13.97	71.94	454	1,362	9,066
chr9q34 (426)	10.73	56.98	359	1,105	3,785
chr18q12 (536)	14.63	72.53	460	1,367	9,198

Table 5.9 Computational time (seconds) for different chromosomes with 20 nearby SNPs algorithm (with the number referred to the number of SNPs in the test chromosome regions)

	chr2 p16	chr2 q37	chr4 q26	chr7 p15	chr7 q21	chr7 q31	chr8 q24	chr9 q34	chr12 q12	chr18 q12
Number	1,050	1,062	1,379	466	1,029	1,112	533	426	665	536
Time	915	918	1,207	393	895	967	454	359	575	460

5.4.2.3 Convergence Rates of the Iterative Algorithm

The following four figures (Figs. 5.21–5.24) show us the convergence properties of the iterative algorithms with different nearby SNPs for the chromosome regions chr7p15, chr8q24, chr9q34 and chr18q12. We can observe that all the iterative algorithms converge fast. Generally, it only needs about 10 iterations for the relative

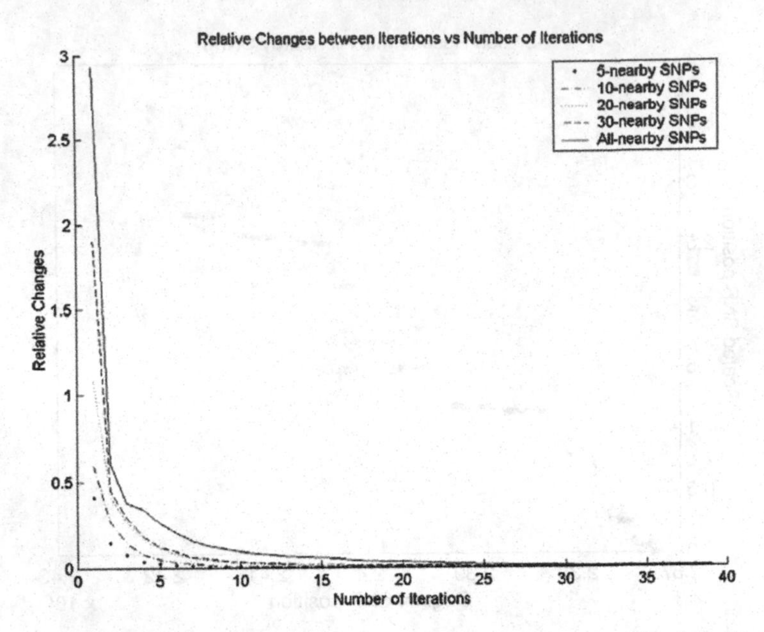

Fig. 5.21 Relative changes between iterations vs. number of the iterations for chr7p15 with the iterative algorithms with the nearby SNP number set at different values

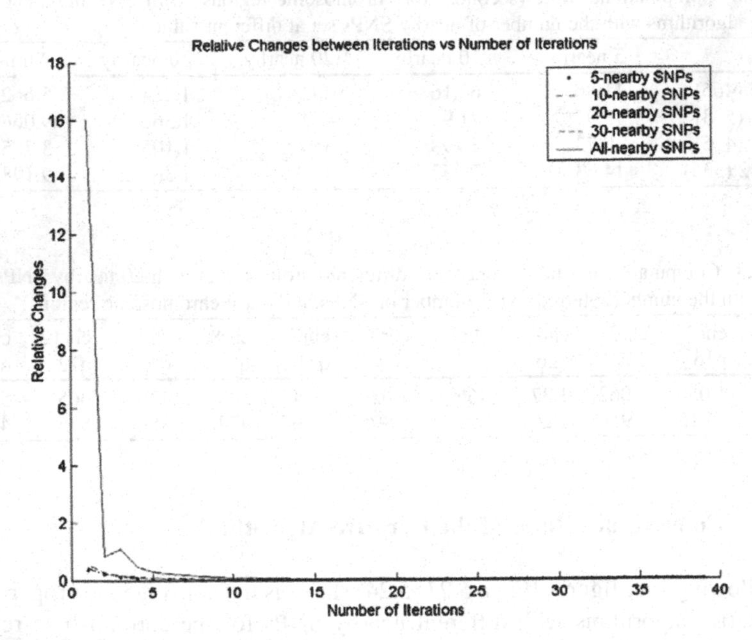

Fig. 5.22 Relative changes between iterations vs. number of the iterations for chr8p24 with the iterative algorithms with the nearby SNP number set at different values

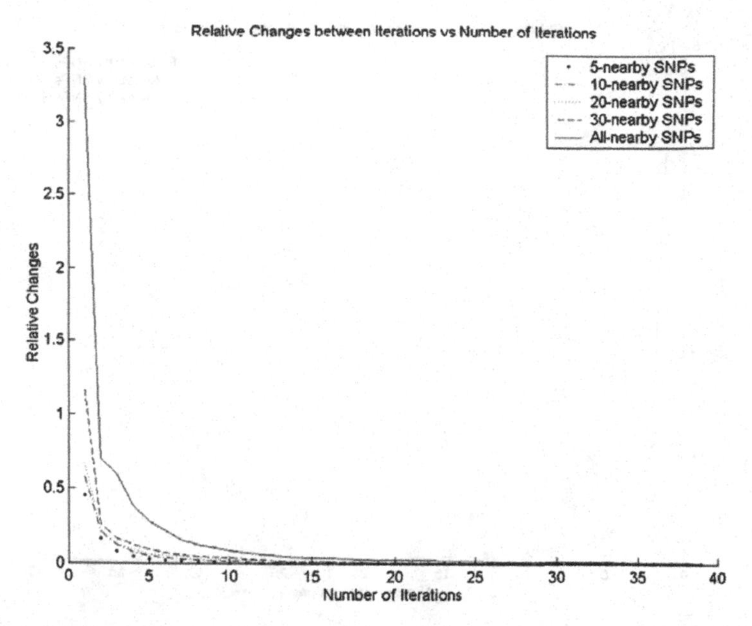

Fig. 5.23 Relative changes between iterations vs. number of the iterations for chr9q34 with the iterative algorithms with the nearby SNP number set at different values

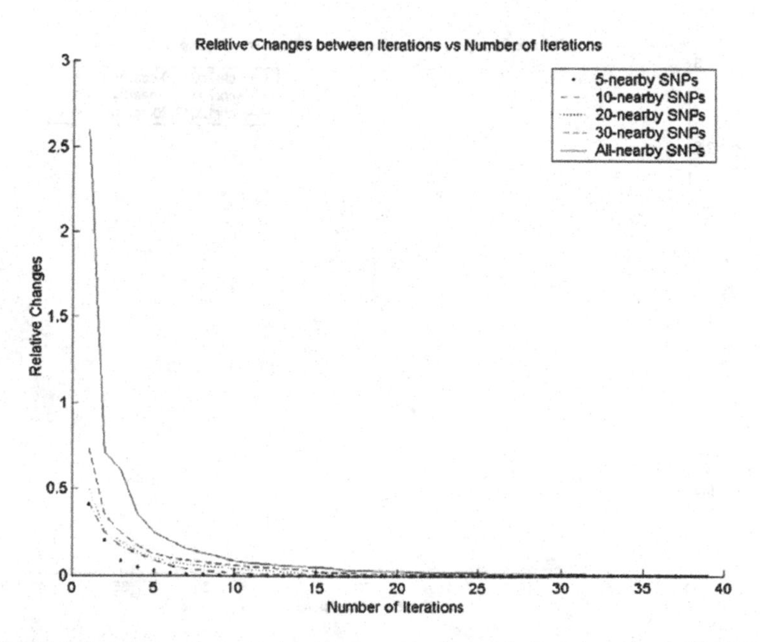

Fig. 5.24 Relative changes between iterations vs. number of the iterations for chr18q12 with the iterative algorithms with the nearby SNP number set at different values

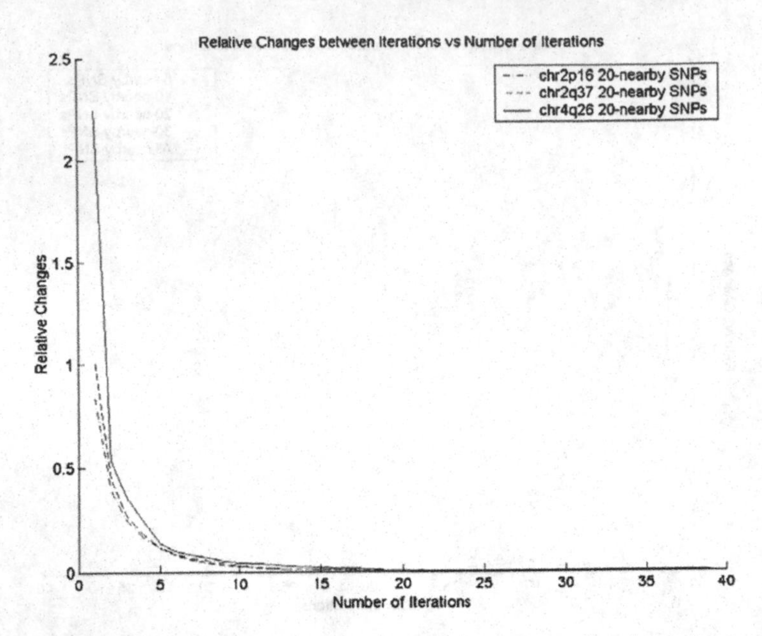

Fig. 5.25 Relative changes between iterations vs. number of the iterations for chr2p16, chr2q37, chr4q26 with the iterative algorithms with 20 nearby SNPs

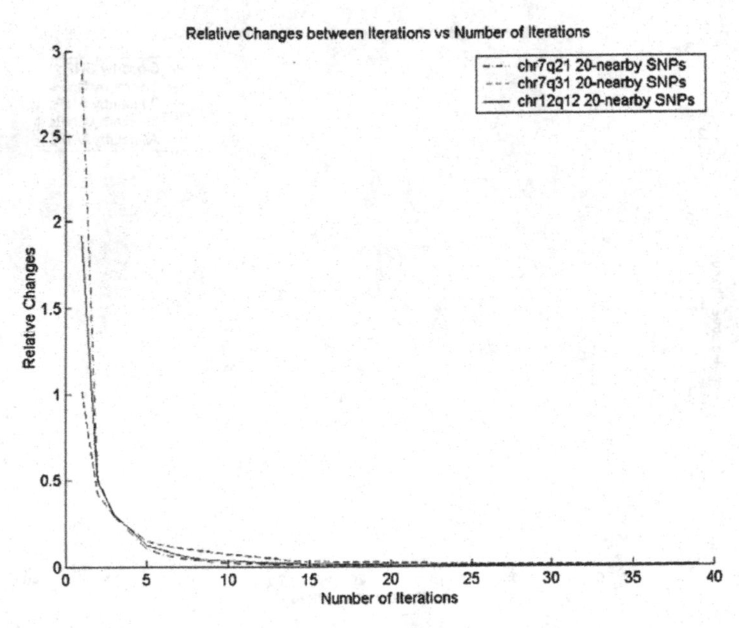

Fig. 5.26 Relative changes between iterations vs. number of the iterations for chr7q21, chr7q31, chr12q12 with the iterative algorithms with 20 nearby SNPs

changes to drop to a level close to zero. It is noted that, with a larger number of nearby SNPs, it usually has a larger initial relative changes.

The two figures (Fig. 5.25 and 5.26) below show us the convergence properties of the remaining six chromosome regions with the iterative algorithm of 20 nearby SNPs. Same as the previous four chromosome regions that we have already seen, the convergence rates are fast and need only about 10 iterations for dropping close to zero.

5.4.2.4 Experimental Results for the Iterative Algorithms with Different Nearby SNPs

The figures below (Figs. 5.27–5.48) show us the scaled SNPs position results with the approximate iterative algorithms of different nearby SNPs. It can be observed that the iterative algorithms with different nearby SNP numbers and the quadratic programming algorithms are producing more or less similar patterns of hot and cold recombination regions. As we have seen previously, the iterative algorithm with all nearby SNPs is observed to produce the nearly the same results. The iterative algorithms with less nearby SNPs are producing results where the hot recombination regions are slightly flatter than those of the all-nearby SNP iterative algorithm and the quadratic programming algorithm.

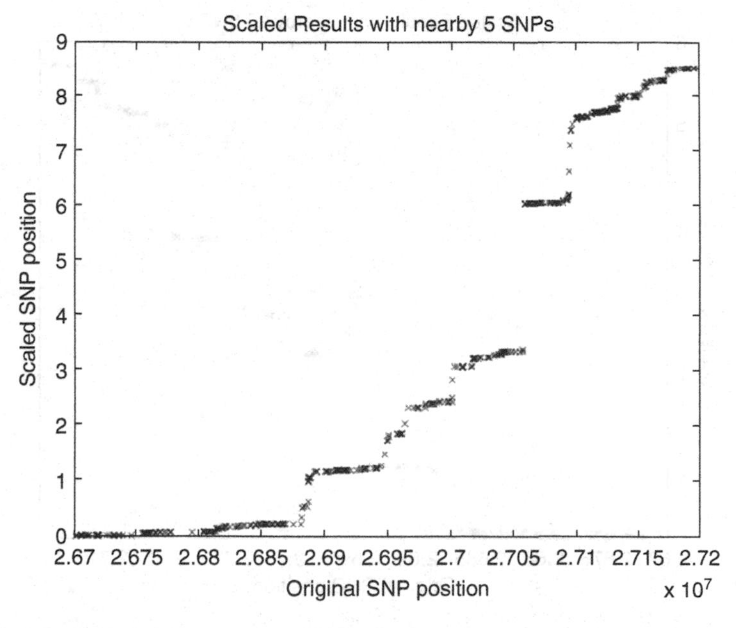

Fig. 5.27 Scaled SNP position vs. original SNP position for chr7p15 with the iterative algorithm of 5 nearby SNPs

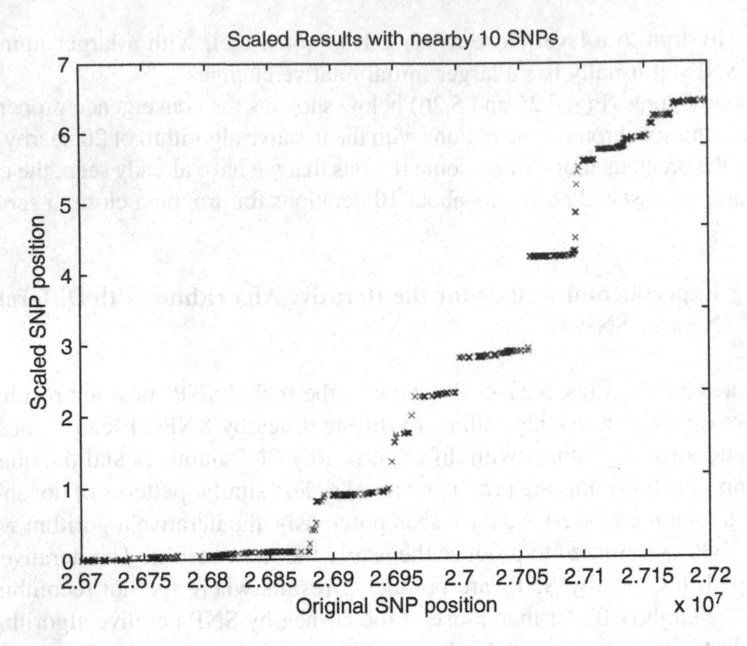

Fig. 5.28 Scaled SNP position vs. original SNP position for chr7p15 with the iterative algorithm of 10 nearby SNPs

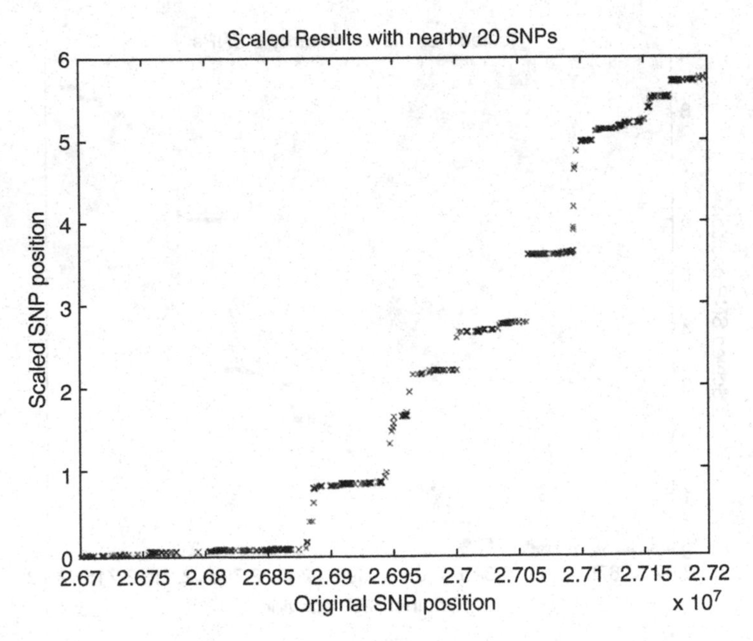

Fig. 5.29 Scaled SNP position vs. original SNP position for chr7p15 with the iterative algorithm of 20 nearby SNPs

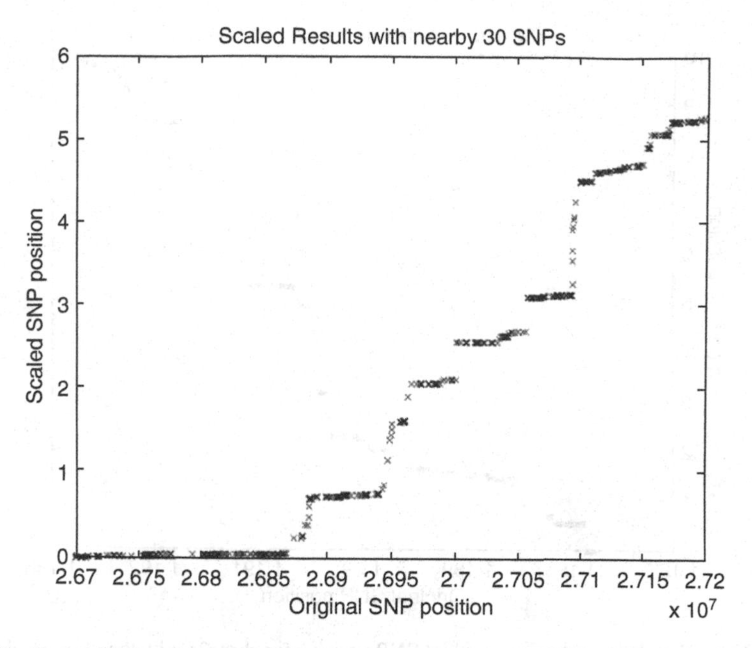

Fig. 5.30 Scaled SNP position vs. original SNP position for chr7p15 with the iterative algorithm of 30 nearby SNPs

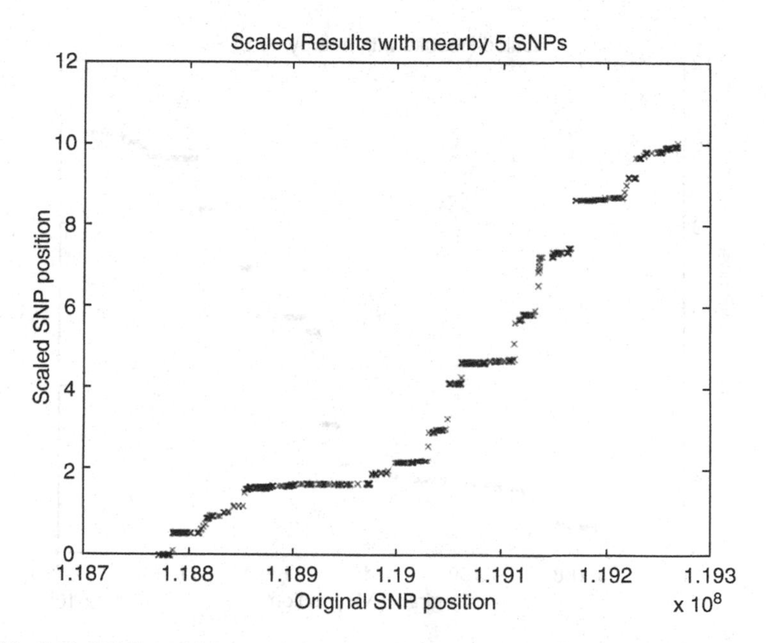

Fig. 5.31 Scaled SNP position vs. original SNP position for chr8q24 with the iterative algorithm of 5 nearby SNPs

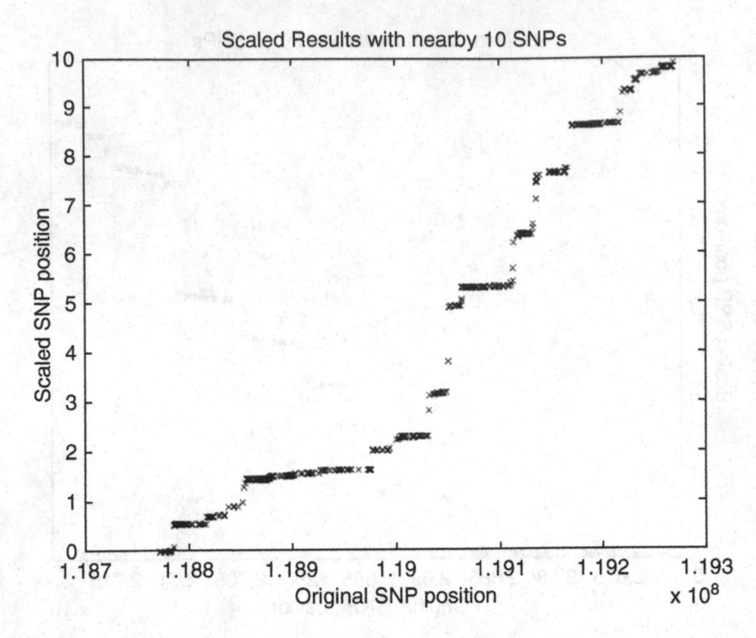

Fig. 5.32 Scaled SNP position vs. original SNP position for chr8q24 with the iterative algorithm of 10 nearby SNPs

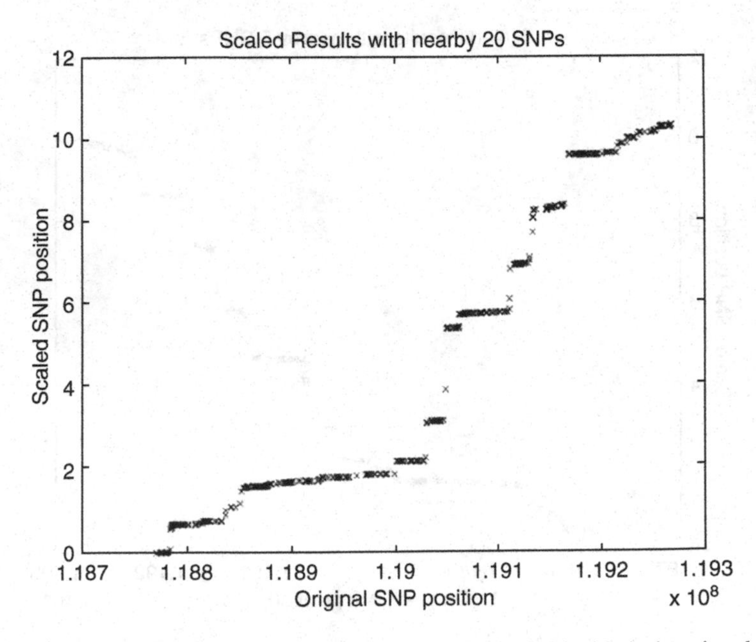

Fig. 5.33 Scaled SNP position vs. original SNP position for chr8q24 with the iterative algorithm of 20 nearby SNPs

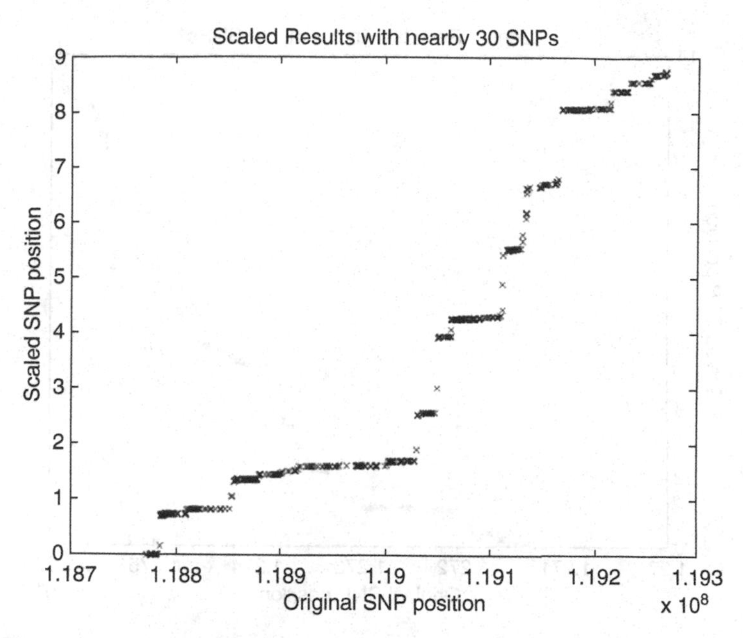

Fig. 5.34 Scaled SNP position vs. original SNP position for chr8q24 with the iterative algorithm of 30 nearby SNPs

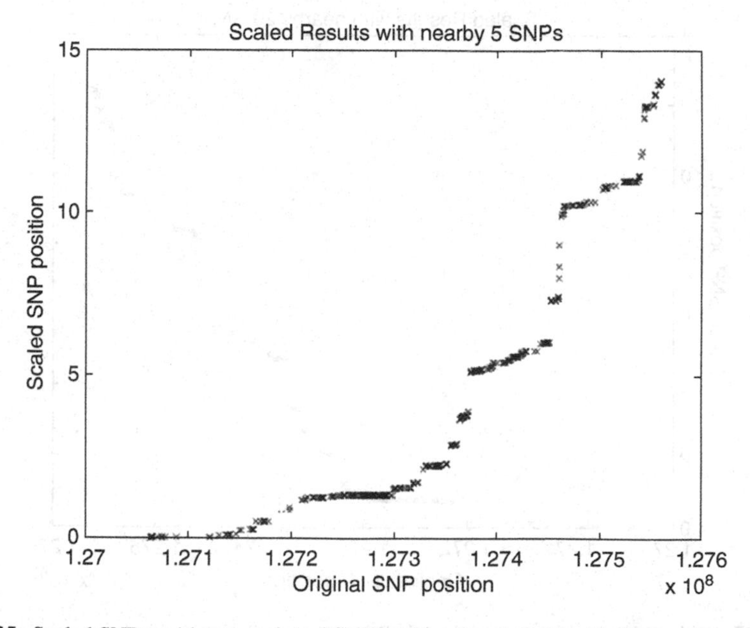

Fig. 5.35 Scaled SNP position vs. original SNP position for chr9q34 with the iterative algorithm of 5 nearby SNPs

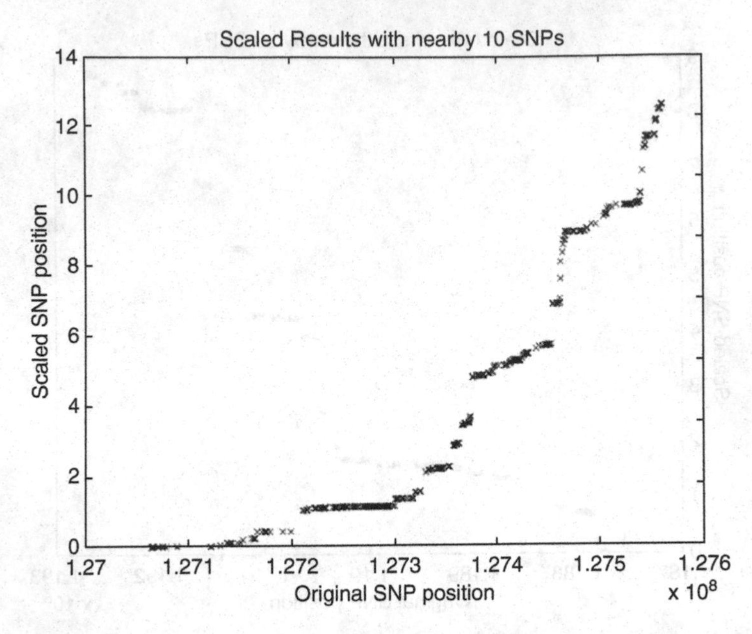

Fig. 5.36 Scaled SNP position vs. original SNP position for chr9q34 with the iterative algorithm of 10 nearby SNPs

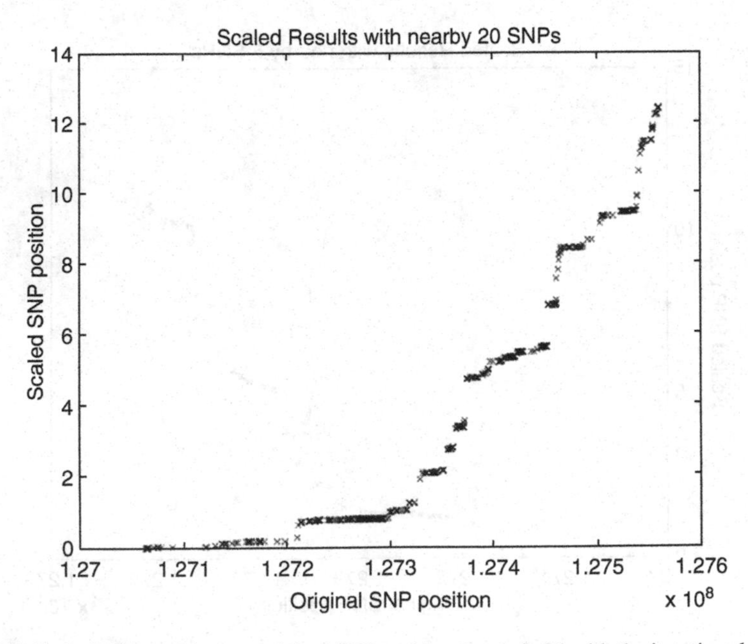

Fig. 5.37 Scaled SNP position vs. original SNP position for chr9q34 with the iterative algorithm of 20 nearby SNPs

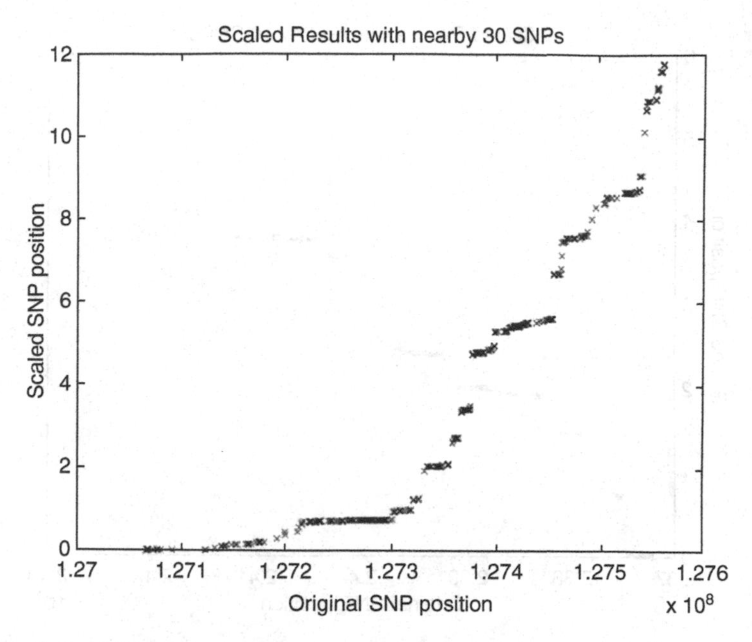

Fig. 5.38 Scaled SNP position vs. original SNP position for chr9q34 with the iterative algorithm of 30 nearby SNPs

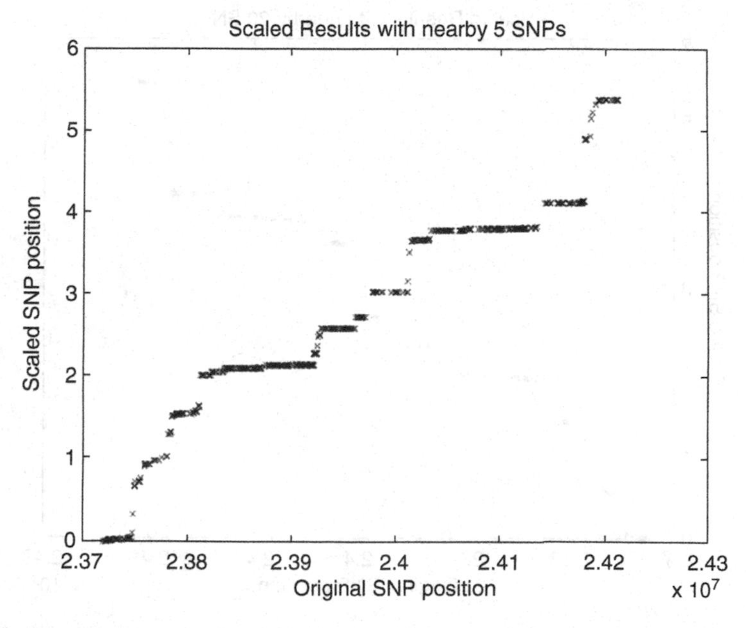

Fig. 5.39 Scaled SNP position vs. original SNP position for chr18q12 with the iterative algorithm of 5 nearby SNPs

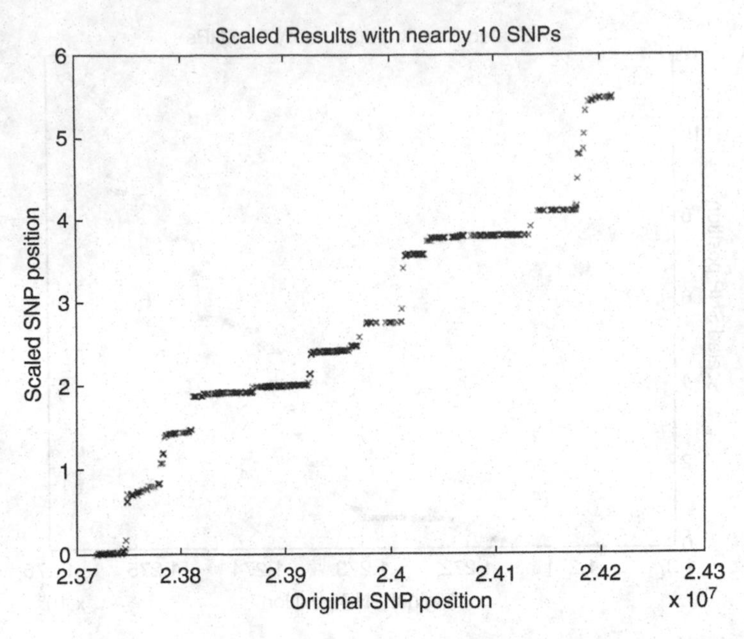

Fig. 5.40 Scaled SNP position vs. original SNP position for chr18q12 with the iterative algorithm of 10 nearby SNPs

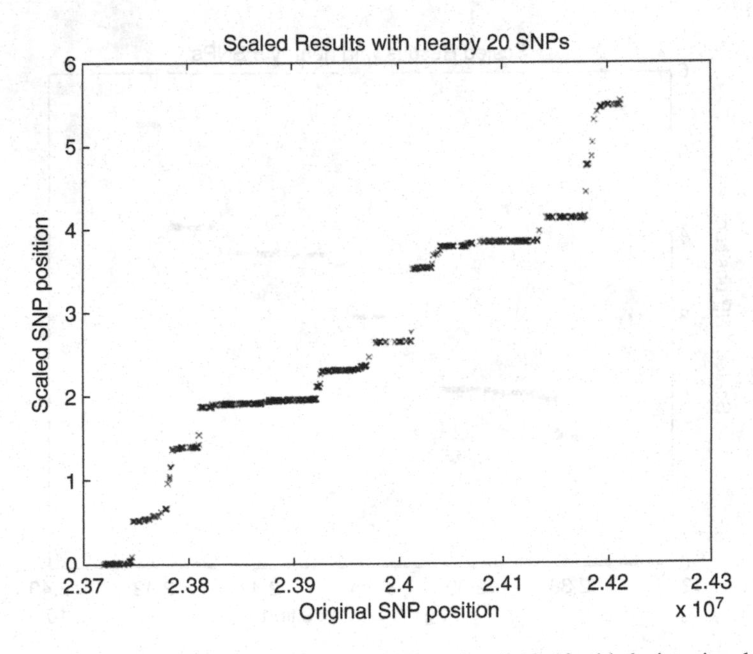

Fig. 5.41 Scaled SNP position vs. original SNP position for chr18q12 with the iterative algorithm of 20 nearby SNPs

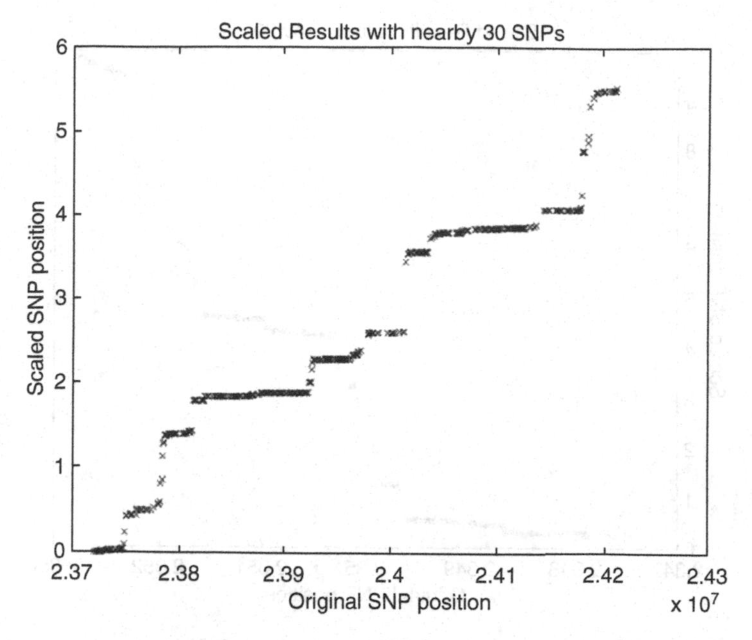

Fig. 5.42 Scaled SNP position vs. original SNP position for chr18q12 with the iterative algorithm of 30 nearby SNPs

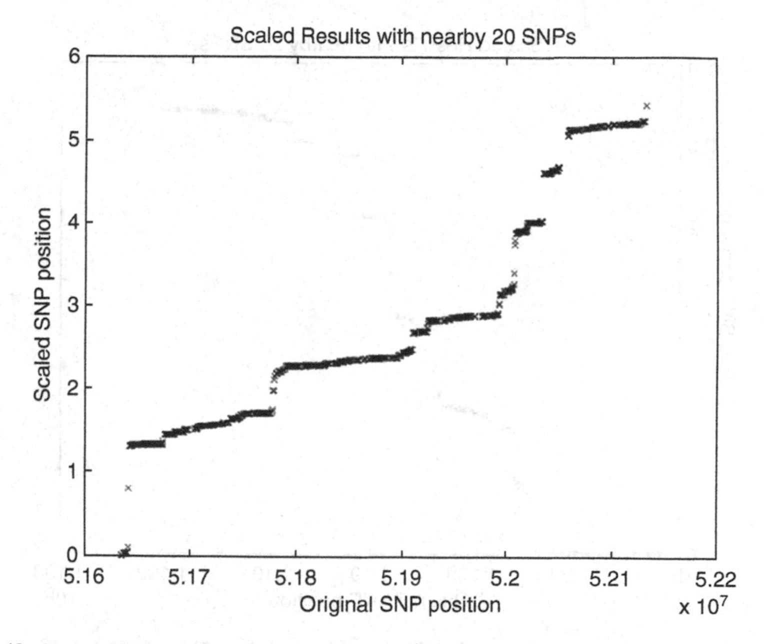

Fig. 5.43 Scaled SNP position vs. original SNP position for chr2p16

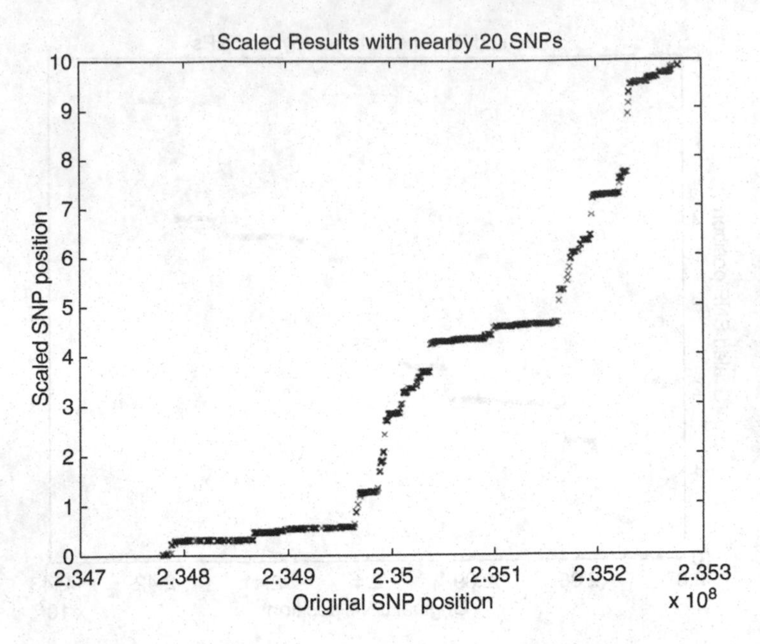

Fig. 5.44 Scaled SNP position vs. original SNP position for chr2q37

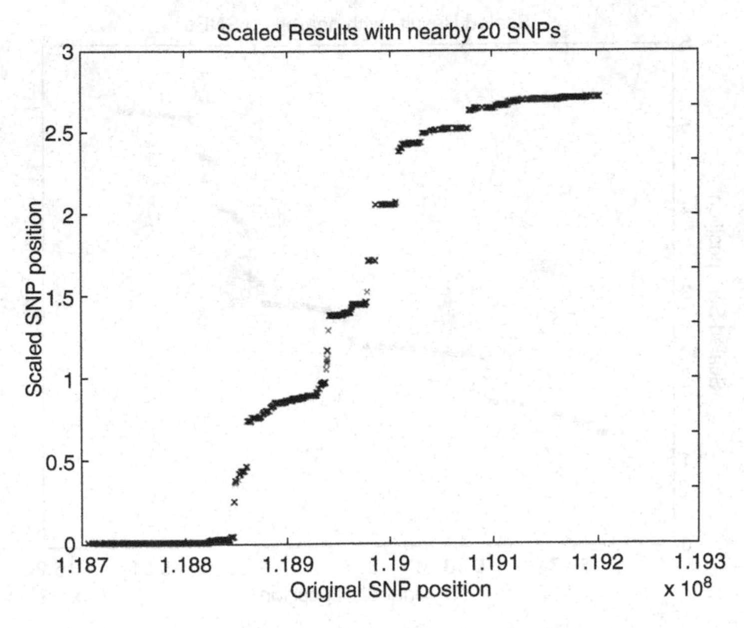

Fig. 5.45 Scaled SNP position vs. original SNP position for chr4q26

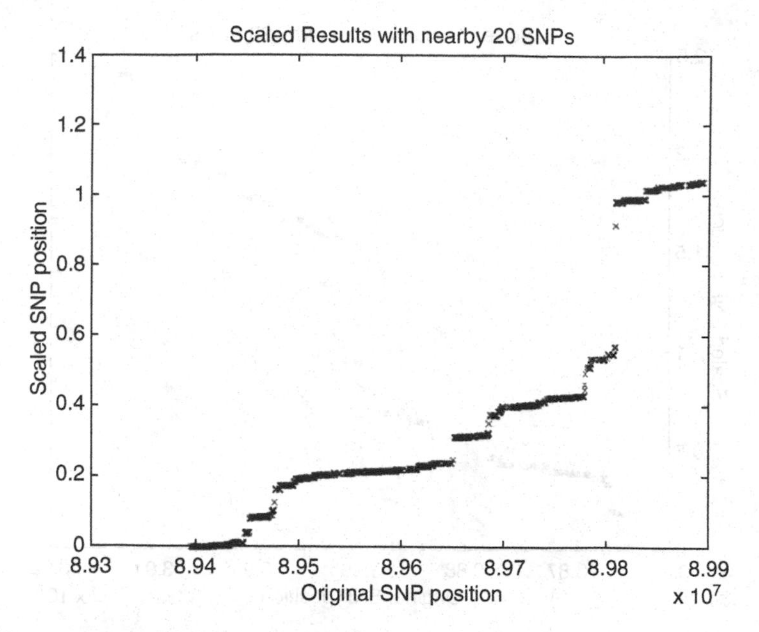

Fig. 5.46 Scaled SNP position vs. original SNP position for chr7q21

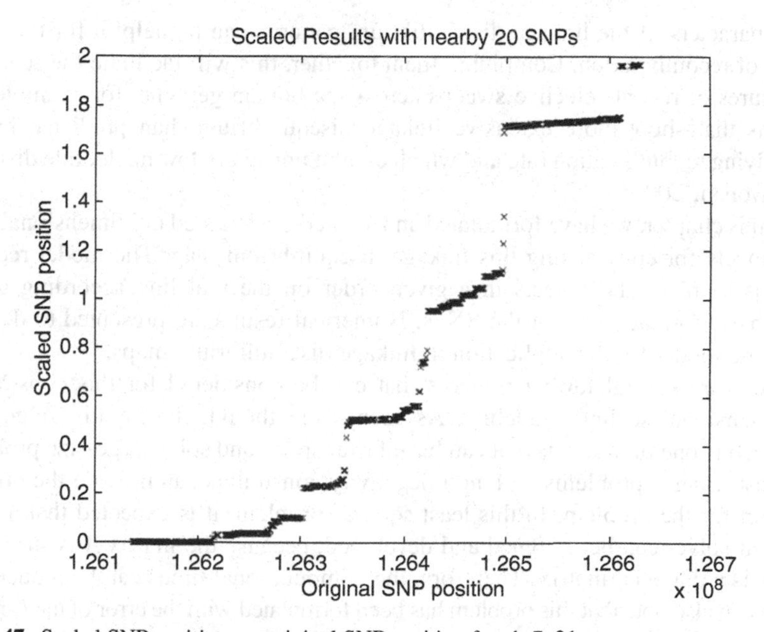

Fig. 5.47 Scaled SNP position vs. original SNP position for chr7q31

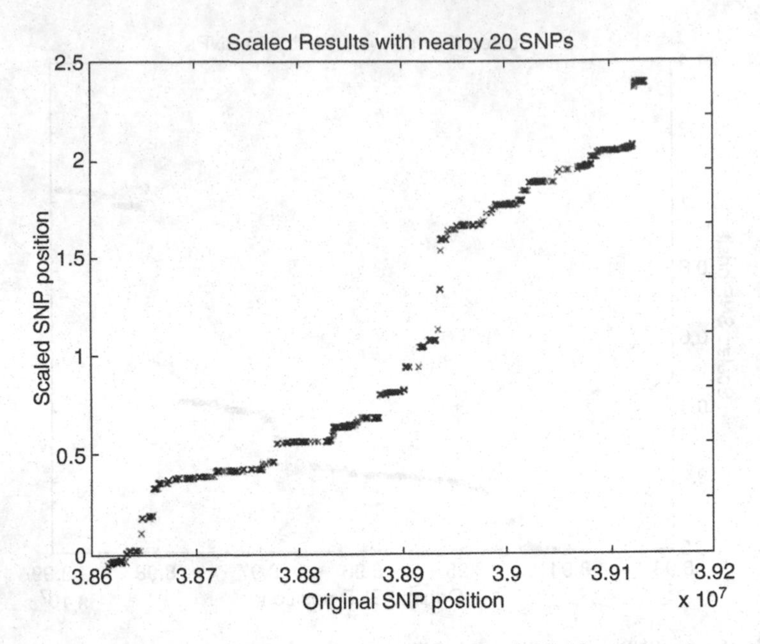

Fig. 5.48 Scaled SNP position vs. original SNP position for chr12q12

5.5 Remarks and Discussions

The characters of the linkage disequilibrium patterns can be helpful for the inferences of recombination. Combining them together, this will facilitate the search of signatures of recent selective sweeps across the human genome, for example, the regions that show more extensive linkage disequilibrium than predicted by the underlying recombination rate and which exhibit unusually low nucleotide diversity (Przeworski, 2002).

In this chapter, we have formulated and studied constrained unidimensional scaling models for constructing this linkage disequilibrium map. The model requirement is to place the objects in a given order on the real line according to the available information about the SNPs. Numerical results are presented to demonstrate the model for the application in linkage disequilibrium maps.

There are several further remarks that can be considered for this constrained unidimensional scaling modeling. As seen, even though the iterative algorithm approach is one of methods that can have faster speed and solve large-size problem, the least squares problems with non-negativity constraints can provide the optimal solution for the problem. In this least squares problem, it is expected that a more efficient solver can be designed and developed, because the matrix A is structured and W is a diagonal matrix. Therefore, the computational times can be reduced.

We can also note that this problem has been formulated with the error of the l_2-norm. Instead, we can further investigate the properties of the error of the other l_p-norm (where $1 \le p < \infty$) for the minimization objective function. Then, the minimization problem will become:

$$J(z_1, z_2, \ldots, z_{n-1}) = \sum_{i>j} w_{ij} \left\| di_j - \sum_{k=j}^{i-1} z_k \right\|_p$$

with the constraints:

$$z_k \ge 0,\ k = 1, 2, \ldots, n-1.$$

For the case of ($p = 1$), the solution for the above problem can be formulated as the solution of a linear programming problem. For the case of ($p > 1$), this becomes a convex programming problem. These mathematical programming problems can be solved by interior point methods.

It may also be interesting to further investigate another variation on this unidimensional scaling when the order of the objects with each of several subsets is known a priori. This information may be available from some previous genomic studies on the chromosome regions concerned or from the medical expert opinions and experience etc.

Chapter 6
Case Study III: Hybrid PCA-NN Algorithms for Continuous Microarray Time Series

In this chapter, we describe about the hybrid models of the principal component analysis (PCA) and neural network (NN) for the continuous microarray gene expression time series. The main contribution of our work is to develop a methodology for modeling numerical gene expression time series. The PCA-NN prediction models are compared with other popular continuous prediction methods. The proposed model can give us the extracted features from the gene expressions time series with higher prediction accuracies. Therefore, the model can help practitioners to gain a better understanding of a cell cycle, and to find the dependency of genes, which is useful for drug discoveries. Based on the results of two public microarray datasets, the PCA-NN method outperforms the other continuous prediction methods. In the time series model, we adapt Akaike's Information Criteria (AIC) tests and cross-validation to select a suitable NN model to avoid the over-parameterized problem.

The outline of this chapter is as followed. In Section 6.1, we describe the background, like the neural network and the transformation algorithms, and their respective applications in the microarray analysis. In Section 6.2, we talk about the motivation for developing the PCA-NN algorithm. In Section 6.3, it is the data description of the public datasets used in our study. In Section 6.4, we describe the details of our proposed methodology and the result comparison of the different methods. In Section 6.5, we discuss about the results of our system and further integration that can be developed, basing on our experimental results.

6.1 Background

6.1.1 Neural Network Algorithms for Microarray Analysis

The function approximation capability of the neural network is one of the network's major properties and advantages. With this property, the researchers can be assured that, provided that appropriate network structure has been employed, the neural network can approximate the real problem accurately. The objective of the function approximate can be formulated as finding function $\Im(.)$ such that:

$$\left\| \Im(x^k) - f(x^k) \right\| < \varepsilon$$

where ε is a small positive number, a set of N different input points are denoted as $\{x^k \in \Re^p, k = 1,2,\ldots,N\}$, a set of N output points $\{d^k \in \Re^q, k = 1,2,\ldots,N\}$, and the actual nonlinear input-output mapping between x and d is denoted as: $f(x^k) = d^k$, where the function $f(.)$ is unknown. The mapping function $\Im:\Re^p \rightarrow \Re^q$ should closely fulfills the relationship: $\Im(x^k) = d^k$, for k = 1, 2,…,N. The property of the function approximation by the neural network is clear with the following theorem of the Universal Approximation Theorem of the network.

Definition: The Universal Approximation Property is said to be satisfied when, with an appropriate number of neurons and optimal weight vector, a neural network can approximate any continuous function, on any compact subset $C \subset \Re^n$ of the input space, to an arbitrary level of accuracy.

Note: A set $S \subset \Re^n$ is called compact if it is closed and bound. A set is closed if and only if its complement in $\Re^n$ is open. A set $S \subset \Re^n$ is open if for every vector $x \in S$, there is an ε-neighbourhood of x: $N(x,\varepsilon) = \{z \in \Re^n | \ \|z - x\| < \varepsilon\}$, such that $N(x,\varepsilon) \in S$. A set is bounded if there is $r > 0$ such that $\|x\| < r$ for all $x \in S$.

Universal Approximation Theorem (Haykin, 1994): Let $\varphi(.)$ denote a bounded and monotone-increasing continuous function. Use I_p to denote the p-dimensional unit hypercube $[0,1]^p$. Let $C(I_p)$ be the space of continuous functions on I_p. For any function $f \in C(I_p)$ and $\varepsilon > 0$, there exists an approximate function $\Im(.)$ that satisfies:

$$\left| \Im(x_1,\ldots,x_p) - f(x_1,\ldots,x_p) \right| < \varepsilon$$

for all $\{x_1,\ldots,x_p\} \in I_p$, where $\Im(.)$ is defined as:

$$\Im(x_1,\ldots,x_p) = \sum_{i=1}^{M} \alpha_i \varphi\left(\sum_{j=1}^{p} w_{ij} x_j - \theta_i \right)$$

for $i = 1, \ldots, M$ and $j = 1, \ldots, p$, where M is an integer, and α_i, θ_i, and ω_{ij} are sets of real constants.

Barron (1991, 1992) estimated that the mean integrated squared error between the target function $f(.)$ and the estimated function &lm;(.) is bounded by:

$$O\left(\frac{C_f^2}{M} \right) + O\left(\frac{M_p}{N} \log N \right)$$

where C_f is the first absolute moment of the Fourier magnitude distribution of the target function $f(.)$, M is the total number of hidden nodes, p is the number of input nodes, and N is the number of training sets.

The multi-layer feedforward network is a member of supervised learning. In supervised learning, a training set (represented by an input vector $\boldsymbol{x}$) and the corresponding desired output vector $\boldsymbol{d}$ are presented to the network. The network is then trained to learn how to minimize the error between its actual output vector $\boldsymbol{z}$ and this desire output vector $\boldsymbol{d}$ (Huang et al., 2004).

The neural network was reported for its successful applications in the gene expression analysis. Herrero et al. (2001) applied neural network for clustering gene expression patterns. Peterson and Ringner (2003) analyzed tumor gene expression profiles with the network. Sawa and Ohno-Machado (2003) developed a neural network-based similarity index for clustering DNA microarray data. Predictions of TP53 gene sequence of values A, C, G and T were modeled with neural network in study (Spicker et al., 2002).

6.1.2 Transformation Algorithms for Microarray Analysis

6.1.2.1 Principal Component Analysis

Among the tools of the dimension reduction and transformation, the principal component analysis (PCA) is a popular tool for many researchers. Its basic idea is to find the directions in the multidimensional vector space that contribute most to the variability of the data. The representation of data by the PCA consists of projecting the data onto the k-dimensional subspace according to

$$x' = F(x) = A^t x$$

where x' is the vectors in the projected space, A^t is the transformation matrix which is formed by the k largest eigenvectors of the data matrix, x is the input data matrix. Let $\{x_1,x_2,\ldots,x_n\}$ be the n samples of the input matrix x. The principal components and the transformation matrix can be obtained by minimizing the following sum of squared error:

$$J_k(a,x') = \sum_{h=1}^{n} \left\| (m + \sum_{i=1}^{k} a_{hi} x_i') - x_h \right\|^2$$

where m is the sample mean, x_i' the i-th largest eigenvector of the co-variance matrix, and a_{hi} the projection of x_h to x_i'.

The principal component analysis was applied to reduce the dimensionality of the gene expression data in studies (Hornquist et al., 2003; Bicciato et al., 2003; Taylor et al., 2002; Yeung and Ruzzo, 2001, etc.). The focuses are on the effective dimensional reduction by the PCA, the analysis of the compressed space and the assistance of the PCA for the classification and the clustering. For example, Hornquist et al. concentrated on the determination of the effective PCA dimensionality. Bicciato et al. described how to use the PCA to reduce the gene expression's dimensional base for a better understanding of its basic biology and to have a better classification result. Taylor et al. applied the PCA to help understand the basis of plant genotype discrimination. Yeung and Ruzzo tested the efficiency of the PCA for clustering gene expression data.

Khan et al. (2001) applied the PCA and neural network for the classification of cancers using gene expression profiling. Khan's purpose is on the classification and the neural network is trained as a classifier for discrete outputs EWS, RMS, BL, and NB of the cancer types. The PCA was employed for the dimensionality reduction of the samples. This can avoid the "over-training" of the network (i.e. low number of parameters as compared to the number of samples). Similarly, we apply the PCA method in our modeling for dimensionality reduction and avoidance of over-fitting.

6.1.2.2 Independent Component Analysis

The independent component analysis (ICA) is a recently developed theory (Hyvärien et al., 2001; Comon, 1994 and Jutten and Herault, 1991). Its objective is to make the transformed entries mutually independent (Theodoridis and Koutroumbas, 2003). Mathematically, let the input samples denoted by x. The task is to determine an N-by-N invertible matrix W, so that the entries $y(i)$, for $i = 0, 1, \ldots, N - 1$, of the transformed vector: $y = Wx$, are mutually independent. Statistically, the requirement for the independence is a stronger condition that the condition of the PCA, which only requires the un-correlation of the components. For Gaussian random variables, these two conditions are equivalent to each other.

The original motivation for the development of the ICA is as followed: Assume that the input data vector x is indeed from a linear combination of statistically independent components. An example is that, of several woofers located in different positions of a room, we have some detectors for checking the sound signals and then for determining the sources from these observed signals. Formally, we have $x = Ay$, where A is known as the mixing matrix of the components y. The task is to determine the de-mixing matrix W as the above paragraph, so that $y = Wx$, for recovering the components of the sources y.

Mathematically, the ICA transformation can work only for non-Gaussian processes, as it is ill-posed for Gaussian processes (Theodoridis and Koutroumbas, 2003). Let the independent components y(i) be all Gaussian. It can be observed that a linear transformation of these components by any unitary matrix will also satisfy the requirement. Thus, PCA should be used in this case, as PCA can return a unique solution by setting a specific orthogonal structure in the transformation. Another condition for the proper working of ICA is that the mixing matrix A must be invertible. For cases where A is a non-square l-by-N matrix, it is required that l must be larger than N and A has to be of full column rank.

The de-mixing matrix W can be estimated by minimizing the mutual information between the transformed random variables. Define the associated entropy of $y(i)$, $H(y(i))$, as (Papoulis, 91):

$$H(y(i)) = -\int p_i(y(i)) \ln p_i(y(i)) dy(i)$$

where $p_i(y(i))$ is the marginal probability distribution function of $y(i)$. Then the mutual information $I(y)$ can be obtained as:

$$I(y) = -H(x) - \ln|\det(W)| + \sum_{i=0}^{N-1} \int p_i(y(i)) \ln p_i(y(i)) dy(i)$$

Techniques of the approximation of this mutual information $I(y)$ and then subsequent minimization of the approximate function can be used for finding the solution of $I(y)$ respective to W (Haykin, 1999; Hyvärien et al., 2001).

6.2 Motivations for the Hybrid PCA-NN Algorithms

Our motivation can be explained by looking at the information transfer mechanism among DNAs, mRNAs and proteins. Genes are used as templates for DNA synthesis. In the transcription process, genes are converted into the messenger RNA (mRNA). While the mRNA is subsequently translated to form proteins, some particular proteins can in turn regulate gene expression profiles. In other words, there exists a complex relationship between the current and future gene expressions values and their lags through this mechanism. Our work is to model this complex time series relationship by using a continuous numerical model. From our algorithm, we can know the influence of each gene on the principal components. With the knowledge of the disease gene, we can apply our algorithm to find out the influential genes in the development of such disease genes. Therefore, the suitable enhancing or inhibiting of the expression of these leading genes could lead to more effective control for the disease gene growth. This can certainly help us to gain a better understanding of genes in a cell cycle, for example, which gene can be understood best for its future analysis.

Our work is different from other studies in bioinformatics, which like Khan's (Khan et al., 2001), have successfully employed the Principal Component Analysis – Neural Network (PCA-NN) as a classifier of gene types. The first step of our proposed PCA-NN system is to form the input vectors for the time series analysis. They are the expression values of the time points in the previous stages of a cell cycle. Then, these input vectors are processed by the PCA. Thirdly, we use these post-processed vectors to feed the neural network predictors. To our best knowledge, our proposed system is the first attempt to employ the principal component analysis and neural network to model the gene expression with continuous input and output values.

One advantage of using neural network is that it can give us continuous modeling output. Khan has used the classification function of the network. Our work can be viewed as an extension of the discrete models to a continuous modeling of the gene expression. The continuous models have the advantage of resembling the real phenomenon better, as the more intricate aspects of gene regulation are dependent not on whether a gene is transcribed but rather on the level of transcription (Wolkenhauer, 2002). In the reverse engineering of the gene profiles, D'haeseleer et al. (1999) have studied the linear model for the network inference that can give us the prediction of the genes' development. The model can be written as:

$$x_i(t+1) = \sum_{j=1}^{J} w_{ij} x_j(t) + b_i$$

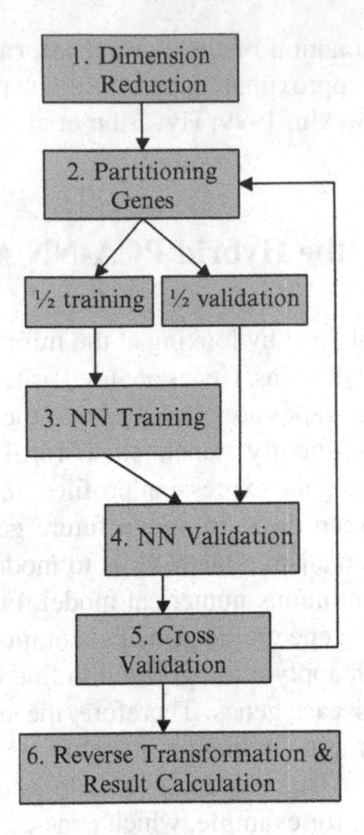

Fig. 6.1 System diagram of the PCA-NN model

where x_i is the expression level of the ith gene, J is the number of gene studied. As pointed out in his research, it is possible to further enhance this model with the neural network. Our modeling can be regarded as a nonlinear generalization of this linear model. The PCA is employed here for feature extraction and we can see the system diagram in Fig. 6.1 below.

6.3 Data Description of Microarray Time Series Datasets

The first dataset is from the experiment of Spellman et al. (1998). It contains the yeast's gene expression levels at different time points of the cell cycle (18 data points in one cell cycle). From the Spellman's data set, there are totally 613 genes that do not have missing values and that show positive cell cycle regulation by periodicity and correlation algorithms. While the number of variables is large, the number of observations per variable is small (18 time points for each gene).

The average absolute percentage change of the genes between two adjacent time points is 94.92%. We will see later that this large volatility of expression levels makes prediction difficult. This value can be proven to be equivalent to the average percentage prediction error of the Naïve method. The Naïve method is one of the most basic while popular methods for time series analysis, and it simply uses the previous realized gene expression value to predict the next coming value. The underlying assumption is that trends and turning points cannot be predicted, and thus, the horizontal line extrapolation is used as the forecast. In other words, the method is equivalent to a random walk model, which shares the same assumption about the structure of the time series data.

We will also test the dataset of Cho et al. (1998). There are 17 time points for a total of 384 genes in this data set. The prediction error for the Naïve method is 28.52%. A potential difficulty with applying the regression-based methods to the microarray data is the possible non-uniformly sampling of the time series data. This can be solved, for example, by regressing the gene expression levels against the various non-uniform time points. Then, extrapolated uniformly sampled time points can be obtained from the derived regression model. Details for the extrapolation method can be found in (Yukalov, 2000, etc.).

6.4 Methods and Results

The first step of our PCA-NN system is to form the input vectors for the time series analysis. They are the expression levels of the time points in the previous stages of the cell cycle. Then, these input vectors are processed by the PCA. Thirdly, we use these post-processed vectors to feed the neural network predictors. Their outputs are compared in the section of result comparison and we can see that the PCA-NN is the most suitable one among the methods we have compared.

6.4.1 Algorithms with Stand-Alone Neural Network

While the gene activities are highly complicated and nonlinear, the neural network is known for its non-linear capability. In this problem, it is used to check if this non-linear method can provide more accurate numerical forecasting. The typical three-layer neural network architecture is employed. The layers are the input layer, the hidden layer and the output layer. The inspiring idea for this structure is to mimic the working of our brain. The above layers correspond to the axons for inputs, synapses, soma, and axons for outputs.

In our computational experiment with the stand-alone neural network, the inputs x_i's will be the expression levels of the lags (of various length) of each gene in turn. And they will make the prediction of the current expression level (denoted by y) of the gene. Mathematically, these inputs x_i's are fed into the neural network structure with the output y as followed (Principe et al., 2000):

$$y = g\left(\sum_{j=1}^{J} w_j^{(2)} f\left(\sum_{i=1}^{I} w_{ji}^{(1)} x_i\right)\right)$$

where I denotes the number of inputs, J the number of hidden neurons, x_i the ith input, $w^{(1)}$ the weights between the input and hidden layers, $w^{(2)}$ the weights between the hidden and output layers.

The lag length of the input variable will be determined by the AIC method, as we will see later. In the PCA-NN experiment, the x_i's will denote the lags of each principal component value in turn. In our study, we have used the tansig activation function:

$$f(x) = \frac{2}{1+e^{-2x}} - 1$$

This is mathematically equivalent to the tanh(x), but runs faster in Matlab than the implementation of the tanh (Vogl et al., 1988). And, we have simply used the linear combination of the inputs for the output activation function.

Our results with the neural network show that the prediction is better than other methods compared but the errors are still high. It may be due to the lack of enough training data, and also due to the fact that the gene expression levels are changing so rapidly that the accurate forecasting is difficult to achieve. The sum of the absolute errors of the prediction is 2,340 while the sum of the absolute gene expression values is 3,921 for the dataset 1. The absolute percentage error is found to be 59.68% for a neural network of ten hidden neurons and a lag length of three previous values. This is still a large percentage error when compared with other time series prediction by the neural network, like the short-term financial time series forecasting. The absolute percentage error for the second dataset is 14.76%. We have studied the effects of using different network architectures and of changing the number of feeding terms for the network. To avoid over-training, we adopt the AIC tests and cross-validation procedure.

6.4.2 Hybrid Algorithms of Principal Component and Neural Network

As the prediction errors made by the stand-alone neural network are still large, the PCA is tested to see if it can assist the neural network for making more accurate prediction. As said earlier, the PCA has been used successfully in the gene expression data analysis to reduce the dimensionality of the data set for better classification results.

We have applied the PCA to the whole spectrum of the genes in the two datasets separately. With the principal components obtained by Singular Value Decomposition (SVD) method, the gene expression matrix of dimension *613* x *18* was reduced to *17* x *18*. In the following Figs. 6.2 and 6.3, the first, second and third principal components are shown. We can see that there exist some more or less clear trends for the neural network to make the prediction. The other major components have similar property too. These principal components will serve as the input vectors for the neural networks. Our purpose of dimension reduction can be said successful.

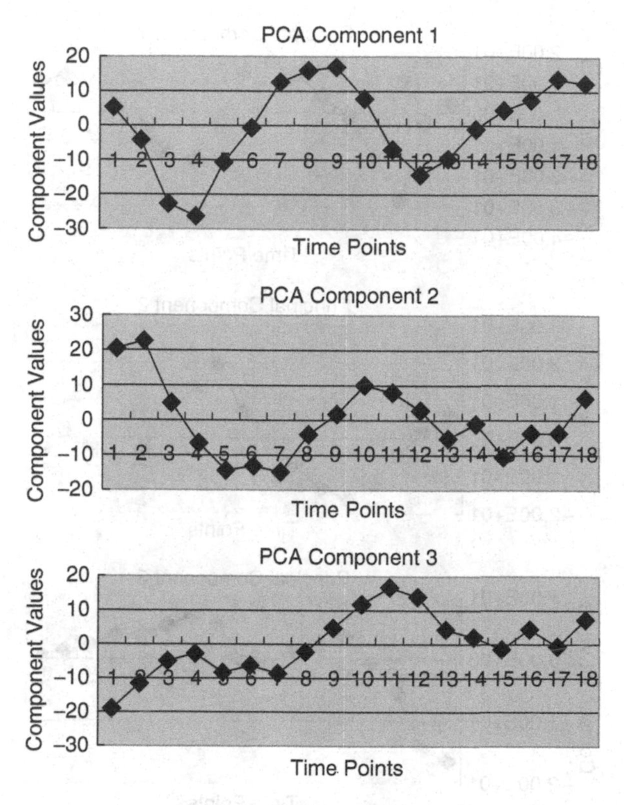

Fig. 6.2 The first, second and third principal components vs. time points for Spellman's dataset

Then, we are going to test its role in the assistance of the neural network's prediction performance. We employ the neural network to make prediction for each principal component. Then, the predictions are transformed back into the original vector space. In the computational experiment with Spellman's dataset, the total absolute prediction error of the neural network in the training (with values at t-1, t-2 and t-3 for predicting the current t-value and with ten hidden neurons) is found to be 1,087. The sum of the gene expression values is 3,921. The absolute percentage error is thus 27.77%, while that for the second dataset is 9.25%.

6.4.3 *Results Comparison: Hybrid PCA-NN Models' Performance and Other Existing Algorithms*

The results from the Naïve prediction, the moving average prediction (MA), which takes the average of past three expression values for prediction, the autoregression (AR(1)), the neural network prediction, the ICA-NN method and the PCA-NN

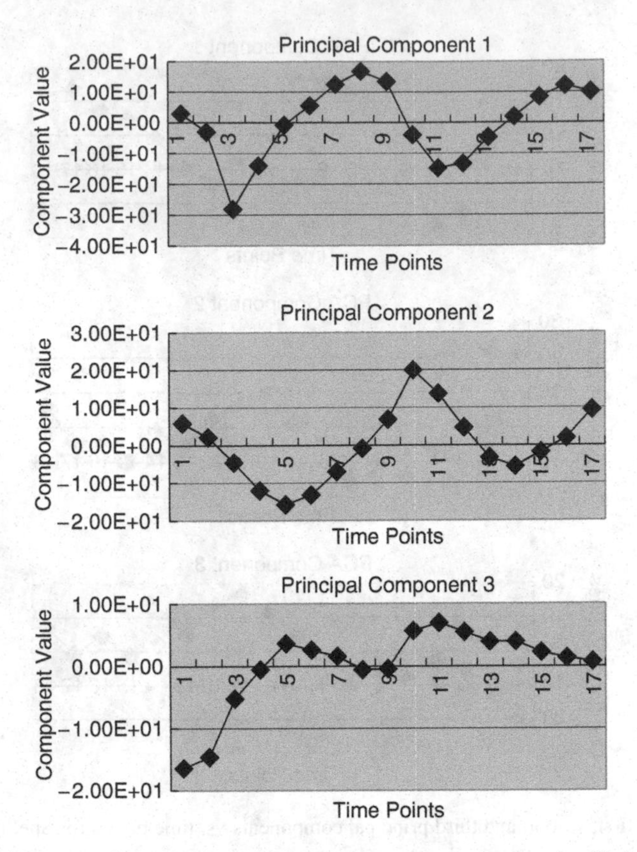

Fig. 6.3 The first, second and third principal components vs. time points for Cho's dataset

Table 6.1 Prediction results from the different methods

Results	Naïve	MA	AR	NN	ICA-NN	PCA-NN
Abs. error (1st set)	94.92%	125.87%	80.34%	75%	83.93%	51.31%
Abs. error (2nd set)	28.52%	39.16%	27.31%	22.52%	27.67%	12.91%

method are listed in Table 6.1. The naïve method simply used the previous expression value as the prediction value. The moving average prediction used the average of a certain number of previous expression values as a predictor. In fact, the Naïve method can be regarded as the moving average of 1-lag model. The first-order autoregression AR(1) is of the form:

$$x_t = \rho x_{t-1} + \varepsilon_t$$

where x_t is the expression level at time t, ρ the coefficient of x_{t-1}. ε_t is the white noise time series with $E[\varepsilon_t] = 0$, $E[\varepsilon_t^2] = \sigma_\varepsilon^2$, and $Cov[\varepsilon_t, \varepsilon_x] = 0$ for all $s \neq t$. These three methods are popular in continuous numerical predictions and their corresponding errors here are the in-sample errors. It can be observed that the NN model and the PCA-NN model are better than these methods. We will see later that the performance of the out-of-sample testing of the PCA-NN model is better than these methods' in-sample testing. The in-sample tests mean that the data is used for both training and testing while the out-of-sample tests mean that the testing data has not been employed in the training process.

6.5 Analysis on the Network Structure and the Out-of-Sample Validations

We tested different combinations of the lag patterns for the training of the neural network with one hidden layer of 10 hidden neurons. Different network architectures were checked for comparing their performances too.

In Table 6.2, the results of feeding the neural network of 10 hidden neurons with different input lag lengths are shown. NN1 represents network with the 1-lag model of value t-1, NN2 with the 2-lag model of values t-1 and t-2, NN3 with inputs t-1, t-2 and t-3, NN4 with input t-1, t-2, t-3 and t-4, NN5 with input t-1, t-2, t-3, t-4 and t-5.

Table 6.3 shows us the performances of the different network architectures. All are of three-layer structure. NN_10 is with 10 hidden neurons, NN_5 with 5 hidden neurons, and NN_20 with 20 hidden neurons.

The AIC results are also listed in Tables 6.2–6.4. It was shown that Akaike's criterion is asymptotically equivalent to the use of cross-validation (Principe et al., 2000). Akaike's information criterion (AIC) is defined as:

Table 6.2 Prediction results with different input lag lengths for stand-alone neural network

Results	NN1	NN2	NN3	NN4	NN5
Abs. error (1st set)	66.23%	63.89%	59.68%	57.82%	55.85%
Abs. error (2nd set)	19.07%	16.68%	14.76%	13.42%	11.77%
AIC (1st set)	189.39	208.31	225.86	245.45	263.63
AIC (2nd set)	375.13	391.73	409.38	425.41	443.31

Table 6.3 Prediction results with different stand-alone neural network structures

Results	NN_5	NN_10	NN_20
Abs. error (1st set)	68.63%	59.68%	47.47%
Abs. error (2nd set)	18.53%	14.76%	10.93%
AIC (1st set)	180.28	225.86	318.77
AIC (2nd set)	364.50	409.38	500.13

Table 6.4 Prediction results for PCA-NN method with different neural network structures

Results	T1N5	T2N5	T3N5	T1N10	T2N10	T3N10
Abs. error (1st set)	67.63%	51.31%	46.85%	57.97%	44.22%	30.99%
Abs. error (2nd set)	22.02%	12.91%	14.65%	19.01%	8.68%	9.47%
AIC (1st set)	160.61	160.87	168.77	184.81	195.75	204.74
AIC (2nd set)	348.91	338.61	351.59	372.42	367.08	389.33

Table 6.5 Prediction results (Abs. percentage error of in-samples and out-of-sample cross-validation) for PCA-NN method with two inputs t-1 and t-2 for neural network structure of 5 hidden neurons

Results	In-1	Out-1	In-2	Out-2
Spellman's dataset	63.48%	70.86%	66.53%	74.04%
Cho's dataset	19.91%	20.18%	16.02%	24.66%

$$AIC = T \ln(residual\ sum\ of\ squares) + 2n \quad (7)$$

where n is the number of parameters estimated, and T the number of usable observations (Enders, 1995). The first term of the above AIC equation is to measure the residual sum of squares, and the second term is a penalty for increasing the number of parameters in the model. While a more parsimonious model has the effect of reducing the residual sum of squares, the AIC test can give us a selection criterion that trades off a reduction in the sum of squares of the residuals for a more parsimonious model. We can use the AIC to aid in the selection of the most appropriate model, which is the model with the smallest AIC value (note that it can be negative).

From the AIC results, the architecture of the t-1 input with 5 hidden neurons is suggested, as it has the smallest AIC value. And, feeding the neural network of 5 hidden neurons with input t-1, the error in the 1st data set is found to be 75% while that in the 2nd data set is 22.52%, which are better than the Naïve, the MA, and the AR methods.

Table 6.4 shows us the prediction results and the AIC values for the PCA-NN method with different network structures. T1 is for feeding the network with input t-1 term only, T2 with input t-1 and t-2 terms, T3 with input t-1, t-1 and t-3 terms. N5 refers to the network of 5 hidden neurons and N10 of 10 neurons. While for the data set 1 the AIC values of TIN5 and T2N5 are more or less the same, the AIC values of the second data set suggest clearly that the model T2N5 should be employed. This AIC result is slightly different from that of the stand-alone neural network.

Table 6.5 lists the results of the cross-validation of the PCA-NN method. The gene expression data is divided into two equal parts. In the first round, In-1 and Out-1 are the prediction errors with the first half as the in-sample data and the second half as out-of-sample data. The In-2 and Out-2 are the results with the opposite partitions. From the table, we can find that the in-sample results are generally better than the out-of-sample results but they are more or less consistent and close to each other. Another observation is that the algorithm's performance of both in-sample and out-of-sample testing is better than other methods' in-sample prediction.

The Fig. 6.4 shows us the poorest prediction results of the Naïve method and the PCA-NN method for Cho's dataset. Among the predictions with the Naïve method, the YDL227c gene expression data has the poorest result, with 65.54% average error. That for the PCA-NN method with the T3N10 model is the gene YHL028W, with 25.45% error. The best prediction result for Naïve method is the gene YDL198c, with 11.62% error. That for PCA-NN prediction result is the gene YBR104w, with only 1.87% error. We can observe that in both cases, the PCA-NN model performs much better that the Naïve method.

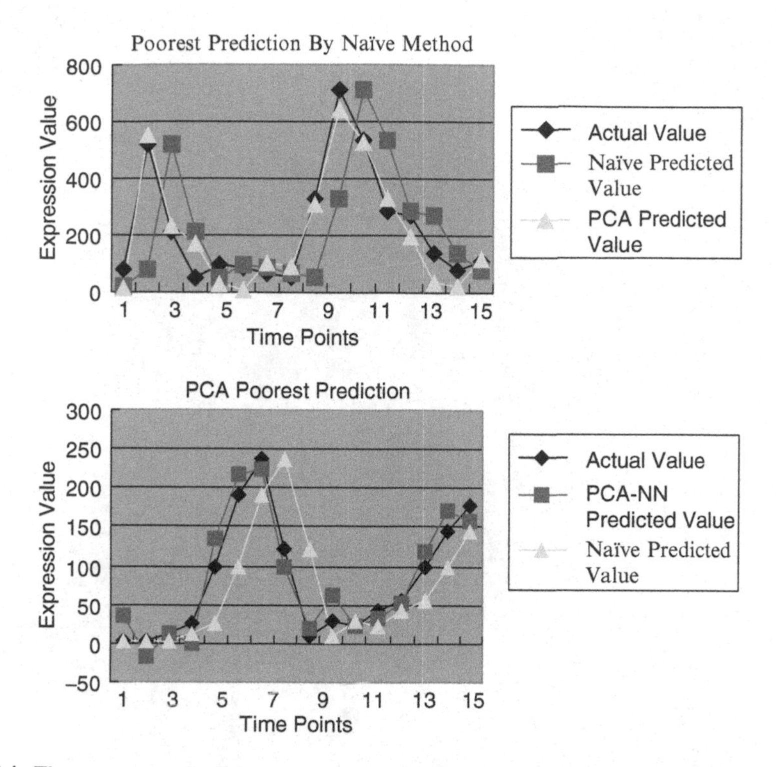

Fig. 6.4 The poorest prediction cases made by Naïve method and PCA–NN method for Cho's dataset (the top one is for the gene YDL227c and the bottom one for the gene YHL028w)

Furthermore, we have tested the possibility of replacing the PCA with the ICA for the modeling. It is interesting to note that the results suggest that ICA does not supplement well with neural network for the gene expression time series modeling here.

6.6 Result Discussions

We study the relationship of the gene expression level in the whole cell cycle from a different prospective. It is to test, given a certain genes expression profile, the possibility of making continuous predictions of the coming gene expression level changes. Our algorithm was applied to two popular gene expression datasets, and it is observed that PCA can assist the network to make more accurate predictions. The PCA-NN predictions have been compared with other popular continuous prediction methods. The results from the Naïve prediction, moving average prediction (MA), autoregression (AR(1)), neural network prediction, ICA-NN method and PCA-NN method are compared. The autoregression results suggest that the stationary time series model is not suitable for this microarray time series problem, while our algorithm is found to model the genomic network more accurately. This is because the genomic network is never truly linear and the neural network is found to be suitable for the gene modeling with continuous outputs.

Chapter 7
Discussions and Future Data Mining Projects

7.1 Tag-SNP Selection and Future Projects

7.1.1 Extension of the CLUSTAG to Work with Band Similarity Matrix

We have seen that, in the CLUSTAG, the clustering and set-cover algorithms are developed to obtain a set of tag SNPs that can represent all the known SNPs in a chromosomal region, subject to the constraint that all SNPs must have a squared correlation $R^2 > C$ with at least one tag SNP, where C is specified by the user. The computational time for the clustering algorithm is of $O(n^2)$. In the computing process, we also need to keep the similarity square matrix between the SNPs pairs.

In order to reduce the computational time and the space requirement, we have tested the implementation of the clustering and set cover methods with the band similarity matrix in our band similarity matrix version of the CLUSTAG. The rational for working with band similarity matrix is that, as we have seen in the discussion in the previous chapters, the correlation between the SNPs will decrease as the distance between them increases. Thus, it is reasonable to ignore the testing of the similarity of SNP pairs when they are far away from each other.

The above four tables (Tables 7.1–7.4) show us the results of the normal similarity matrix (full) and the band similarity matrix (band). The number of clusters formed by these two kinds of matrix is more or less the same. We can observe that the number of tag SNPs in both cases are more or less the same, with the full matrix case requiring one less tag SNP in the sample chromosome regions. Nevertheless, as we have said, the computational time for the band matrix is much less than the full matrix case. It can be observed that, with the band r^2 matrix, the computational time is faster than full matrix by about 25%. The theoretical computation time for the band matrix is $O(n)$, while that for the full matrix is $O(n^2)$ as we have seen.

Table 7.1 Tagging results of chromosome 9 of full similarity matrix vs. band similarity matrix (band width = 100 kb, Total SNPs = 440)

	Complete	Minimax	Graph
Tag SNP (full)	132	117	117
Tag SNP (band)	133	117	117

Table 7.2 Tagging results of chromosome 18 of full similarity matrix vs. band similarity matrix (band width = 100 kb, Total SNPs = 545)

	Complete	Minimax	Graph
Tag SNP (full)	124	115	115
Tag SNP (band)	125	116	116

Table 7.3 Tagging results of chromosome 8 of full similarity matrix vs. band similarity matrix (band width = 100 kb, Total SNPs = 539)

	Complete	Minimax	Graph
Tag SNP (full)	135	124	124
Tag SNP (band)	136	126	126

Table 7.4 Computational time for the CLUSTAG minimax algorithm (in seconds)

	Chromosome 8	Chromosome 9	Chromosome 18
Run time (full)	25.078	22.516	18.281
Run time (minimax)	17.750	16.782	13.703

7.1.2 Potential Haplotype Tagging with the CLUSTAG

Our preliminary results below have shown support that the CLUSTAG can be extended to the clustering of the tag haplotypes instead of the tag SNPs. By working with haplotypes, we are extending the considerations of pair-wise SNPs in each step of the CLUSTAG to the considerations between multi-loci alleles (haplotypes) in each step. The comparison of working with multi-loci and single-locus alleles has been discussed previously.

From the test data of the segment of the chr3p from 47,000,000 to 47,100,000, there exist 26 SNPs. The haplotypes in this region can be found with the EM algorithm etc. The correlation of these 325 haplotypes can be found with the methods like the Haploview. This pre-processing step has been done by the expert from the HKU's Genome Research Centre. These pairs of correlations between the haplotypes are fed into the CLUSTAG. Again, the threshold is set to 0.8 and we have found that a total of seven tag haplotypes are enough for tagging these haplotypes. The Fig. 7.1 below shows us the graphical result of the tagging haplotypes.

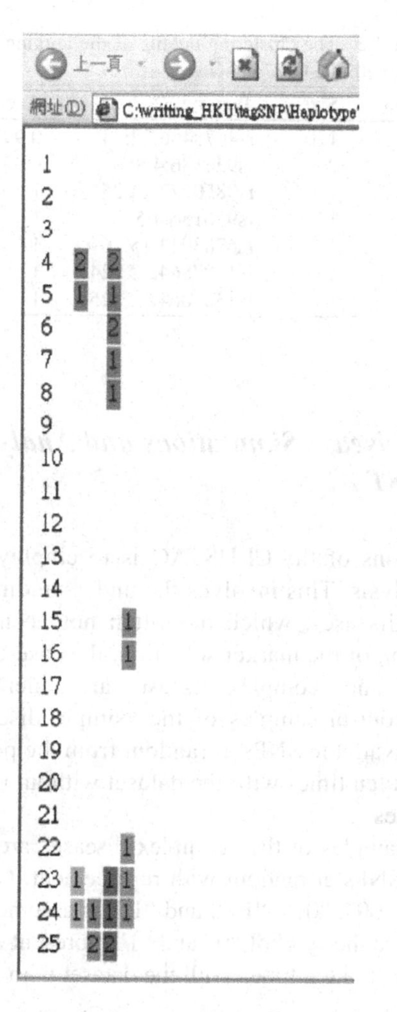

Fig. 7.1 The results for the tagging haplotypes with the CLUSTAG

In the above Table 7.5, the names of the haplotypes contain the following information:

1. The id of the first SNP
2. The relative position of the first SNP in the set of markers
3. The relative position of the last SNP in the set of markers

It can be observed that the ratio of the tagged haplotypes to the tag haplotypes is very high. This is because there is a lot of overlapping between the haplotypes so that the correlations between these haplotypes are very high. On the other hand, with only 26 SNP markers, there are 325 haplotypes estimated. Thus, there will be a computational problem if the number of SNPs increases, and skills from the parallel computing etc. may be needed for solving this problem.

Table 7.5 The clustering details of the tagging haplotypes with the CLUSTAG

Cluster	Size	Tagging SNP	Avg. sim.
0	120	rs4078466:22:24	0.997483
1	200	rs936186:4:8	0.993985
2	1	rs9850277:24:25	1
3	1	rs936186:4:5	1
4	1	rs6783943:15:16	1
5	1	rs13078642:23:24	1
6	1	rs13078642:23:25	1

7.1.3 *Complex Disease Simulations and Analysis with CLUSTAG*

One of the applications of the CLUSTAG is to employ the tag SNPs for the complex disease analysis. This involves the understanding of the genetic architecture of complex diseases, which has often not been fully understood yet. Thus, in the evaluation of the marker selection, the case-control samples of both the "simple disease" and "complex disease" are generated by the simulation methods. The case-control samples of the "simple disease" are simulated by selecting one of the available SNPs at random from the population data set. This procedure is repeated ten times with the dataset without replacement to simulate the ten simple diseases.

The case-control samples of the "complex disease" are simulated by selecting two of the available SNPs at random with replacement. One of the four different combination patterns "00", "01", "10", and "11" is assumed to be the cause of the complex disease, where the symbol "0" and "1" represent the two different alleles. This procedure is repeated ten times with the dataset with replacement to simulate the ten simple diseases.

The data set for the simulated disease locus study is from the population CEU on Chr18:23717221..23737220 (build: NCBI B34, downloaded on 26 April, 2006). The LD data dump is from the direct HapMap LD output. This dataset has the LD information of 56 SNPs. There are 90 person samples in the dataset. The result with the minimax algorithm has 17 tag SNPs. The compression ratio is 17/56 = 30.36%, and the compactness of the clusters is 0.0016.

Three popular methods of disease analysis-the logistic regression, neural network and decision tree, have been applied to the simulated data sets. The comparison results are listed in the tables below (Tables 7.6–7.9). The misclassification rates are the average values for the ten "simple diseases" and the ten "complex diseases". From these results, it can be observed that:

1. The logistic regression is more efficient for the analysis of the "simple diseases".
2. The neural network is more efficient for the analysis of the "complex diseases".

Table 7.6 Experimental results of "simple disease" with 60% training and 40% validation

Misclassification rate	Logistic regression	Neural network	Decision tree
Training (orig.)	0.0019	0.0037	0.0056
Validation (orig.)	0	0.0083	0.0083
Training (tag SNP)	0.0019	0.0056	0.0056
Validation (tag SNP)	0	0.0028	0.0083

Table 7.7 Experimental results of "complex disease" with 60% training and 40% validation

Misclassification rate	Logistic regression	Neural network	Decision tree
Training (orig.)	0.0389	0.0019	0.0260
Validation (orig.)	0.0444	0.0028	0.0306
Training (tag SNP)	0.0463	0.0167	0.0259
Validation (tag SNP)	0.0500	0.0111	0.0306

Table 7.8 Experimental results of "simple disease" with 40% training and 60% validation

Misclassification rate	Logistic regression	Neural network	Decision tree
Training (orig.)	0.0028	0.0167	0.0028
Validation (orig.)	0	0.0074	0.0074
Testing (orig.)	0	0.0037	0.0037
Training (tag SNP)	0.0028	0.0083	0.0038
Validation (tag SNP)	0	0.0296	0.0074
Testing (tag SNP)	0	0.0370	0.0038

Table 7.9 Experimental results of "complex disease" with 40% training and 60% validation

Misclassification rate	Logistic regression	Neural network	Decision tree
Training (orig.)	0.0333	0.0194	0.0333
Validation (orig.)	0.0333	0.0074	0.0333
Testing (orig.)	0.0519	0.0259	0.0519
Training (tag SNP)	0.0361	0.0167	0.0399
Validation (tag SNP)	0.0407	0.0296	0.0481
Testing (tag SNP)	0.0519	0.0593	0.0519

3. The minimax algorithm of the program CLUSTAG can effectively reduce the number of the SNPs genotyped, while the misclassification rates with the experiments with the tag SNPs are still at low level.
4. The experiment results are relatively stable for the different partition ratios.

It would be interesting to further investigate the prediction results of these models with the CLUSTAG and WCLUSTAG for more complicated simulated disease models (Thornton-Wells et al., 2006).

7.2 Algorithms for Non-parametric LD Maps Constructions

7.2.1 Localization of Disease Locus with LD Maps

Maniatis et al. (2005) applied their parametric LD Map for testing an 890kb region flanking the *CYP2D6* gene associate with poor drug-metabolizing activity. This is for refining the location of a causal mutation. With the LD map, they succeeded in locating the functional polymorphism at 14.9kb from its true location with the result shown in Chapter 5. In Chapter 5, we have also shown that our quadratic programming approach can produce similar LD map. We observed that the basic shapes of both figures are similar with each other. The cold-spot regions and the hot-spot regions occur in similar regions of the chromosome. It is expected that the location of disease locus may also be achieved with the LD map of the quadratic programming approach.

A simple model with the scaled LD map for the location of disease is as followed. Let there be N SNPs in one chromosome region, and their scaled LD distances have been computed to be $S_1, S_2, \ldots, S_n$ with ascending order. Assume that a disease D is caused by the mutation in one allele within this chromosome region. Similar with our scaled LD model, let w_{iD} be a positive weighting parameter that reflects the accuracy of the dissimilarity d_{iD} between the SNP i and the disease D. The key issue is to identify a scaled position X, starting from the scaled position of the first SNP S_1, such that X can best approximate the observed dissimilarities between the SNP objects and the disease. Mathematically, the problem is to minimize the objective function:

$$J(X) = \sum_{i=1}^{k} w_{iD}\,[X - (S_i + d_{iD})]^2 + \sum_{i=k+1}^{n} w_{iD}\,[X - (S_i - d_{iD})]^2$$

where X is assumed to be contained within the scaled region S_k to S_{k+1}. Similarly, for each value of k from 1 to n, we can obtain one minimum value of X for the $J(X)$. The minimum of these $J(X)$'s is the scaled location of the X. This outlines a simple model for the localization of disease with our scaled LD map. More sophisticated models of localization of disease with our scaled LD map can be further investigated for the result comparison. Some research scientists are investigating this problem of the location of disease with our scaled method, and their computational results show that the localization of disease can also be achieved with our scaled LD map.

7.2.2 Other Future Projects

Besides the further investigation works that we have discussed above and in the discussion section of the chapter on non-parametric LD maps, we can also consider the different means of locating the hot recombination regions and cold recombination regions. For example, we can consider the possibility of applying the sliding window approach for locating the hot recombination and cold recombination regions mathematically. It is noted previously that the hot and cold recombination regions can be identified with our graphical outputs of the scaled SNP position.

For finding the hot recombination region (chr9q34) with moving windows of 2 SNP intervals (3 SNPs), we have found that the position with the maximum scaled LD distance between consecutive SNPs for the quadratic programming algorithm is with the starting position at 214. It means that the interval position is from 214 to 215. We can look at the names of these SNPs and locate their genetic location. The genetic interval for this interval is from 127,373,454 to 127,374,341. Similarly, the position with the maximum scaled LD distance between consecutive SNPs for the iterative algorithm with 20 nearby SNPs is also at 214. The interval position is from 214 to 215. And this is the same as the quadratic programming algorithm. We can see from the figure (Fig. 7.2) below that there are cases that these regions do not overlapping exactly. This is not strange as we have pointed out previously that the approximate iterative algorithms with different nearby SNPs usually produce LD maps with less sharpness in the hot recombination regions.127,294,178.

For locating the cold recombination region (chr9q34), with moving windows of 100 SNP intervals (101 SNPs), we have found that the position with the minimum

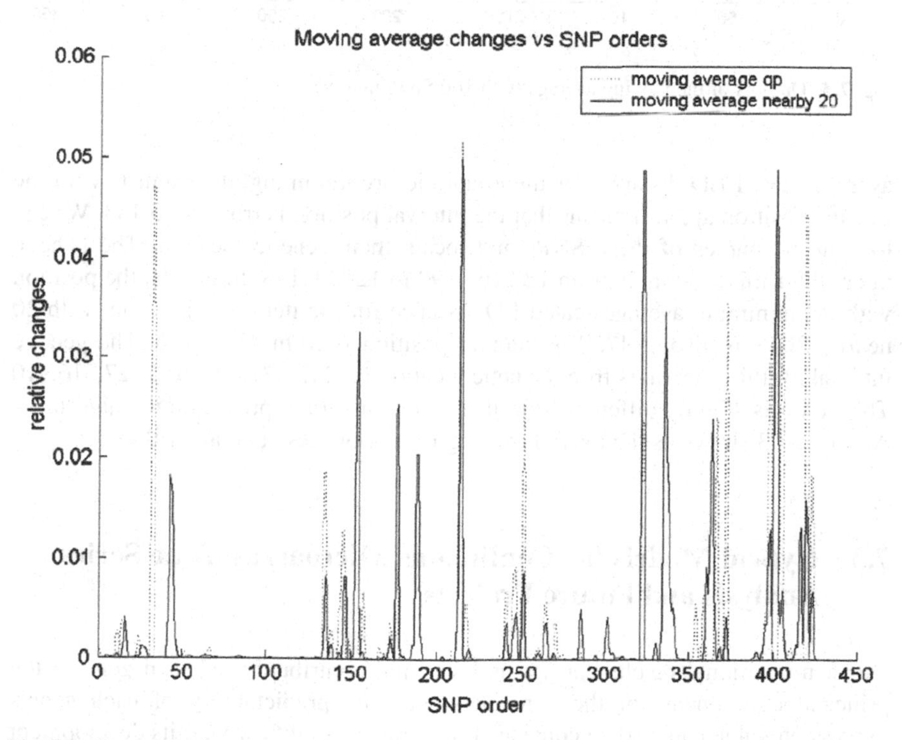

Fig. 7.2 Outputs of the moving window with 2 SNP intervals

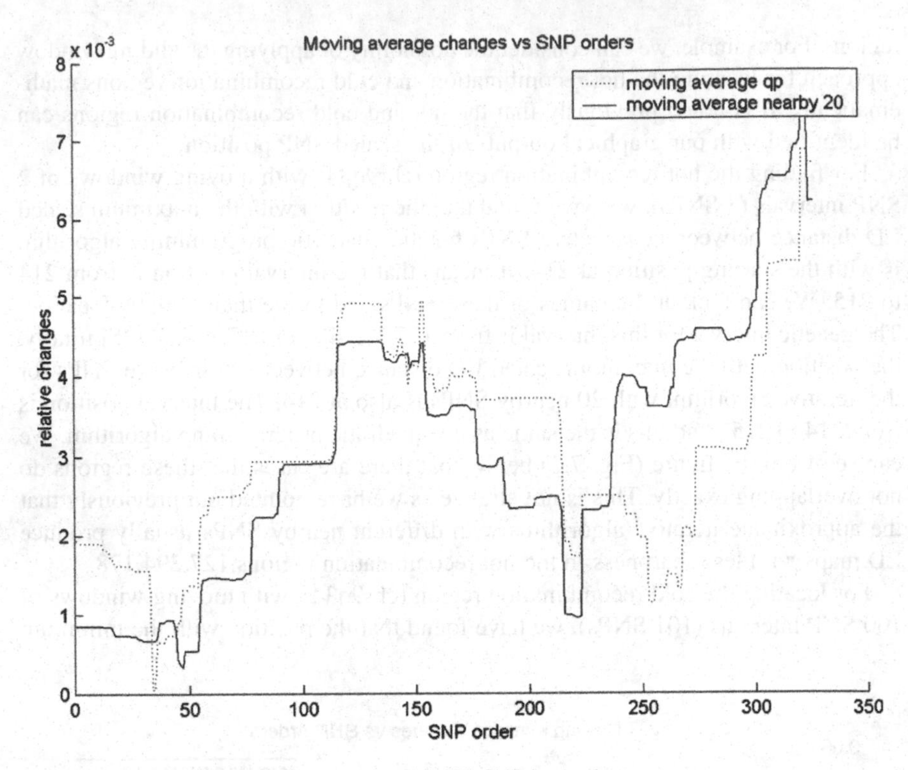

Fig. 7.3 Outputs of the moving average with 100 SNP intervals

average scaled LD distance for the quadratic programming algorithm is with the starting position at 34. It means that the interval position is from 34 to 133. We can look at the names of these SNPs and locate their genetic location. The genetic interval for this interval is from 127,167,190 to 127294178. Similarly, the position with the minimum average scaled LD distance for the iterative algorithm with 20 nearby SNPs is also at 47. The interval position is from 47 to 146. The genetic interval for this interval is from Genetic location = 12127,215,076o 12127,316,910 This result is slightly different from that of the quadratic programming algorithm. And Fig. 7.3 shows us the overall moving results of these two algorithms.

7.3 Hybrid Models for Continuous Microarray Time Series Analysis and Future Projects

Our computational results can let us know the contribution of each gene to the principal components of the gene network. The predictability of each gene's expression value can also be considered as a measure of how well its development can be understood. It is because we have considered the time series data of the gene expression in its whole life cycle. A good prediction model means that we can identify the correct principal component for influencing the gene's developments.

The neural network has been known for its non-linear function capability. Its prediction error is quite reasonable, which is better than the other methods like the Naïve method and the AR method. From the results of the two popular gene expression datasets, we can see that the PCA can assist the neural network to make more accurate predictions and the PCA-NN method outperforms others. A main difficulty in our numeral prediction is that the time points in one cell cycle are short. The changes of the expression levels are very large between each time interval. In short, we need to do further work on this short multivariable time series analysis of the yeast's cell cycle in order to further improve the prediction results. Our system can also be seen as a nonlinear gene inference network. It can give us more accurate model of the genome network, which is never truly linear, while a large-scale gene expression predictive model can obviate the need for an exact understanding of the system at the biochemical level (D'haeseleer et al., 1999).

Genetic algorithm is a promising tool for the optimization of the gene weightings as pointed by Keedwell and Narayanan (2002). Similarly, we can regard our NN numerical prediction as the fitness function of the GA. We are going to select the most influential genes for each gene's development in its life cycle with the GA. The experimental results here have clearly shown that our proposed PCA-NN outperforms the other methods of linear regression, simple neural network and ICA-NN etc. Thus, the suitable candidate to work with the GA will be the PCA-NN model, forming the hybrid GA-PCA-NN system. Another potential method is the ensemble learning, which have been successfully applied for the classification problems in microarray (for example, Tan and Gilbert, 2003). Our goal is to achieve a nonlinear gene network that can utilize the microarray data fully, with continuous inputs and continuous outputs, and that can provide us the details of the genes' developmental dependencies. This can be helpful for drug development of the enhancing or inhibiting of a specific gene.

Bibliography

Aach, J. and Church, G. 2001. Aligning gene expression time series with time warping algorithms. Bioinformatics, 17(6), 495–508.

Abecasis, G. et al. 2001. Extent and distribution of linkage disequilibrium in three genomic regions. American Journal of Human Genetics, 68, 191–197.

Acta, A. 2001. Chemometric strategies for normalisation of gene expression data obtained from cDNA microarrays. Analytica Chimica Acta, 446(1–2), 449–464.

Adams, M. D. et al. 1995. Initial assessment of human gene diversity and expression patterns based upon 83 million nucleotides of cDNA sequence. Nature, 377(Suppl.), 3–174.

Ahmadian, A. et al. 2000. Single-nucleotide polymorphism analysis by Pyrosequencing. Analytical Biochemistry, 280, 103–110.

Ahuja, R., Magnanti, T., and Orlin, J. 1993. Network flows-theory, algorithms, and applications. Englewood Cliffs, NJ: Prentice Hall.

Akdemir, B. 2008. Ensemble adaptive network-based fuzzy inference system with weighted arithmetical mean and application to diagnosis of optic nerve disease from visual-evoked potential signals. Artificial Intelligence Medicine, 43(2), 141–149.

Alberts, B. et al. 1994. Molecular biology of the cell. 3rd Ed. New York: Garland.

Alderborn, A. et al. 2000. Determination of single nucleotide polymorphisms by real-time pyrophosphate DNA sequencing. Genome Research, 10, 1249–1258.

Altshuler D. et al. 2000. An SNP map of the human genome generated by reduced representation shortgun sequencing. Nature, 407, 513–516.

Amaratunga, D. and Cabrera, J. 2004. Exploration and analysis of DNA microarray and protein array data. New York: Wiley.

Angluin, D. 1992. Computational learning theory: survey and selected bibliography. In Proceedings of the Twenty-Fourth Annual ACM Symposium on Theory of Computing, 351–369.

Antonellis A. et al. 2002. A method for developing high-density SNP maps and its application at the Type 1 Angiotensin II receptor (AGTR1) locus. Genomics, 79, 326–332.

Ao, S. 2003a. Automating stock prediction with neural network and evolutionary Computation', Intelligent Data Engineering and Automated Learning. In Proceedings of the Fourth International Conference on Intelligent Data Engineering and Automated Learning 2003, Hong Kong, March 2003. Pages 203–210. Springer.

Ao, S. 2003b. Using fuzzy rules for prediction in tourist industry with uncertainty. In Proceedings of the Fourth International Symposium on Uncertainty Modeling and Analysis, University of Maryland, College Park, MD, USA, September 21–24, 2003. Pages 213–218. IEEE.

Ao, S. 2003c. Hybrid intelligent system for pricing the indices of dual-listing stock markets. In Proceedings of the IEEE/WIC International Conference on Intelligent Agent Technology, Halifax, Canada, October 13–17, 2003, pp.495–498. IEEE.

Ao, S. 2006. A framework for neural network to make business forecasting with hybrid VAR and GA components. Engineering Letters (International Association of Engineers), 13(1), 24–29.

Ao, S. 2007. Neural network regressions with fuzzy clustering. In Proceedings of the 2007 International Conference of Information Engineering of World Congress on Engineering 2007, London, UK, July 2–4, 2007, pp. 507–512. ISBN: 978-988-98671-5-7.

Ao, S. 2008. Constructing linkage disequilibrium map with iterative approach. In Current Themes in Engineering Technologies: World Congress on Engineering and Computer Science. American Institute of Physics.

Ao, S. and Ng, M. 2006. Gene expression time series modeling with principal component and neural network. Soft Computing – A Fusion of Foundations, Methodologies and Applications, 10(4), 351–359.

Ao, S., Ng, M., and Ching, W. 2004. Modeling gene expression network with PCA-NN on continuous inputs and outputs basis. In Current Trends in High Performance Computing and Its Applications. Proceedings of the High Performance Computing and Applications 2004, Shanghai, China, August 8–10 2004. Pages 209–214.

Ao, S., Ng, M., and Sham, P. 2005a. Constrained unidimensional scaling. In Programme and Abstracts, 3rd World Conference on Computational Statistics & Data Analysis, International Association for Statistical Computing, pp. 49.

Ao, S., Yip, K., et al. 2005b. CLUSTAG: hierarchical clustering and graph methods for selecting tag SNPs. Bioinformatics, 21(8), 1735–1736.

Ao, S., Ng, M., and Sham, P. 2007. Constrained unidimensional scaling with application to genomics. Computational Statistics & Data Analysis. The Official Journal of the International Association for Statistical Computing, 52(1), 201–210.

Ao, S., Amouzegar, M., and Chen, S. (Eds.). 2008. Current Themes in Engineering Technologies: World Congress on Engineering and Computer Science. American Institute of Physics.

Arfken, G. 1985. Mathematical methods for physicists, 3rd Ed. Orlando, FL: Academic.

Bakker, P. et al. 2005. Efficiency and power in genetic association studies. Nature Genetics, 37, 1217–1223.

Barnes, M. R. and Gray, I. C. (Eds.). 2003. Bioinformatics for geneticists. New York: Wiley.

Barrett, J. et al. 2005. Haploview: analysis and visualization of LD and haplotype maps. Bioinformatics, 21(2), 263–265.

Barron, A. R. 1991. Complexity regularization with application to artificial neural networks. In Nonparametric functional estimation and related topics, G. Roussas (Ed.). Boston, MA/ Dordrecht: Kluwer, 561–576.

Barron, A. R. 1992. Neural net approximation. Proceedings of the Seventh Yale Workshop on Adaptive and Learning Systems. New Haven, CT: Yale University, 69–72.

Barnsley M. F. 1988. Fractals everywhere. Boston, MA: Academic.

Beineke, L. and Wilson, R. 1997. Graph connections-relationships between graph theory and other areas of mathematics. Oxford/New York: Oxford University Press.

Bergeron, B. 2003. Bioinformatics computing. Upper Saddle River, NJ: Prentice Hall.

Berry, M. and Linoff, G. 2004. Data mining techniques: for marketing, sales, and customer relationship management. New York: Wiley.

Bicciato et al. 2003. PCA disjoint models for multiclass cancer analysis using gene expression data. Bioinformatics, 19(5), 571–578.

Biswas, S., Storey, J., and Akey, J. 2008 Mapping gene expression quantitative trait loci by singular value decomposition and independent component analysis. BMC Bioinformatics, 9(1), 244.

Bodmer, W. and Bonilla, C. 2008. Common and rare variants in multifactorial susceptibility to common diseases. Nature Genetics, 40, 495–701.

Boggs, P. and Tolle, J. 1995. Sequential quadratic programming. In Acta numerica, A. Iserles (Ed.). Cambridge: Cambridge University Press, pp. 1–51.

Boguski, M. S. and Schuler, G. D. 1995. Establishing a human transcript map. Nature Genetics, 10, 369–371.

Boser B., Guyon I., and Vapnik V. 1992. A training algorithm for optimal margin classifiers. In Proceedings of the Annual Conference on Computational Learning Theory, ACM Press, Pittsburgh, PA, pp. 144–152.

Bosl, W. 2007. Systems biology by the rules: hybrid intelligent systems for pathway modeling and discovery. BMC Systems Biology, 1, 13, doi: 10.1186/1752-0509-1-13.

Brookes, K. et al. 2006. The analysis of 51 genes in DSM-IV combined type attention deficit hyperactivity disorder: association signals in DRD4, DAT1 and 16 other genes. Molecular Psychiatry. Advance online publication. 8 August 2006, pp.1–20.

Brown, M. et al. 2000. Knowledge-based analysis of microarray gene expression data by using support vector machines. PNAS, 97(1), 262–267.

Buetow, K. et al. 2001. High-throughput development and characterization of a genomewide collection of gene-based single nucleotide polymorphism markers by chip-based Matrix-assisted laser desorption/ionization time-of-flight mass spectrometry. PNAS USA, 98, 581–584.

Butte, A. et al. 2001. Comparing the similarity of time-series gene expression using signal processing metrics. Journal of Biomedical Informatics, 34, 396–405.

Byng, M. et al. 2003. SNP subset selection for genetic association studies. Annals of Human Genetics, 67, 543–556.

Byrne, C. 2008 . Sequential unconstrained minimization algorithms for constrained optimization. Inverse Problems, 24, doi:10.1088/0266-5611/24/1/015013.

Cai C., Han L., Ji Z., and Chen Y. 2004. Enzyme family classification by support vector machines. Proteins, 55, 66–76.

Carlson, C. et al. 2004. Selecting a maximally informative set of single-nucleotide polymorphisms for association analyses using linkage disequilibrium. American Journal of Human Genetics, 74, 106–120.

Castillo, O., Xu, L., and Ao, S. (Eds.). 2008. Trends in intelligent systems and computer engineering. New York: Springer.

Causton, H. et al. 2003. Microarray gene expression data analysis: a beginner's guide. Oxford: Blackwell.

Chambers, J. et al. 2008 Common genetic variation near MC4R is associated with waist circumference and insulin resistance. Nature Genetics, 40, 716–718.

Chen, T., Filkov, V., and Skiena, S. 2001. Identifying gene regulatory networks from experimental data. Parallel Computing, 27, 141–162.

Cherry, J., et al. 1997. Genetic and physical maps of Saccharomyces cerevisiate. Nature, 387, 67–73.

Cho, J. 2008. The genetics and immunopathogenesis of inflammatory bowel disease. Nature Reviews Immunology, 8, 458–466.

Cho, R. et al 1998. A genome-wide transcriptional analysis of the mitotic cell cycle. Molecular Cell, 2, July 1998. 65–73.

Cho Y. et al. 2004. Multifactor-dimensionality reduction shows a two-locus interaction associated with Type 2 diabetes mellitus. Diabetologia, 47(3), 549–554.

CIGMR. 2005 (modified date: March 22 2005). Tagging SNPs. Web Address: http://slack.ser.man.ac.uk/theory/tagging.html.

Clark, A. et al. 2005. Ascertainment bias in studies of human genome-wide polymorphism. Genome Research, 15, 1496–1502.

Clayton D. 2001. http://www.nature.com/ng/journal/v29/n2/extref/ng1001-233-S10.pdf.

Coffey C. et al. 2004 An application of conditional logistic regression and multifactor dimensionality reduction for detecting gene-gene interactions on risk of myocardial infarction: the importance of model validation. BMC Bioinformatics, 5, 49.

Collins A., Lonjou, C., and Morton, N. E. 1999. Genetic epidemiology of single-nucleotide polymorphisms. Proceedings of the National Academy of Science, 96, 15173–15177.

Collins, A., Lau, W., and Vega, F. 2004. Mapping genes for common diseases: the case for genetic (LD) maps. Human Heredity, 58, 2–9.

Comon, P. 1994. Independent component analysis-a new concept? Signal Processing., 36, 287–314.

Costa, I.G., et al. 2002. A symbolic approach to gene expression time series analysis. Neural Networks 2002 Brazilian Symposium, pp. 25–30.

Couzin, J. 2002. New mapping projects splits the community. Science, 296, 1391–1393.

Couzin, J. and Kaiser, J. 2007. Genome-wide association: closing the net on common disease genes. Science, 316(5826), 820–822.

Cowles, C., Joel, N., Altshuler, D., and Lander, E. 2002. Detection of regulatory variation in mouse genes. Nature Genetics, 32, 432–437.

Craig, P., Kennedy, J., and Cumming, A. 2002. Towards visualising temporal features in large scale microarray time-series data. Information Visualisation, 2002. Proceedings. Sixth International Conference on, 10–12 July 2002. 427–433.

Daly, M. et al. 2001. High-resolution haplotype structure in the human genome. Nature Genetics, 29:2, 229–232.

Datta S. and Datta S. 2003. Comparisons and validation of statistical clustering techniques for microarray gene expression data. Bioinformatics, 19, 459–466.

Dawson, K. 2000. The decay of linkage disequilibrium under random union of gametes: how to calculate Bennett's principal components. Theoretical Population Biology, 58, 1–20.

Dawson, E. et al. 2002. A first-generation linkage disequilibrium map of human chromosome 22. Nature, 418, 544–548.

Deerwester, S. et al. 1990. Indexing by latent semantic analysis. Journal of the American Society for Information Science, 41(6), 391–407.

De Martinville. B. et al. 1982. Assignment of first random restriction fragment length polymorphism (RFLP) locus (D14S1) to a region of human chromosome 14. American Journal of Human Genetics, 34, 216–226.

DeRisi, J., et al. 1996. Use of a cDNA microarray to analyse gene expression patterns in human cancer. Nature Genetics, 14, 457–460.

Dewey, T. 2002. From microarrays to networks: mining expression time series. Information Biotechnology Supplement. Drug Discovery Today, 7(20), 170–175.

D'haeseleer, P., Liang, S., and Somogyi, R. 1999. Gene expression data analysis and modeling. Pacific Symposium on Biocomputing.

Ellis, T. et al. 1998. Chemical cleavage of mismatch: a new look at an established method/recent developments. Human Mutation, 11, 345–353.

Enders, W. 1995. Applied Econometric Time Series. New York: Wiley.

Escoffier, L. 2001. Analysis of population subdivision. In Handbook of statistical genetics. Balding, D. et al.(Eds.). New York: Wiley.

Esposito , F. et al. 2008. Independent component model of the default-mode brain function: combining individual-level and population-level analyses in resting-state fMRI. Magnetic Resonance Imaging, doi:10.1016/j.mri.2008.01.045.

Fearnhead , P. and Donnelly, P. 2001. Estimating recombination rates from population genetic data. Genetics, 159, 1299–1318.

Finch, H. 2005. Comparison of distance measures in cluster analysis with dichotomous data. Journal of Data Science, 3, 85–100.

Fisher, S. et al. 2008. Genetic determinants of ulcerative colitis include the ECM1 locus and five loci implicated in Crohn's disease. Nature Genetics, 40, 710–712.

Foulds, L. 1991. Graph theory applications. New York: Springer.

Fujito T. 2001. On approximability of the independent/connected edge dominating set problems. Information Processing Letters, 79, 261–266.

Futschik, M. and Kasabov, N. 2002. Fuzzy clustering of gene expression data. Fuzzy Systems, 2002. FUZZ-IEEE'02. In Proceedings of the 2002 IEEE International Conference on 12–17 May 2002, 1, 414–419.

Gabriel, S., Schaffner, S., Nguyen, H., Moore, J., Roy, J., Blumenstiel, B., Higgins, J., DeFelice, M., Lochner, A., Faggart, M., et al. 2002. The structure of haplotype blocks in the human genome. Science, 296, 2225–2229.

Ganapathiraju, M., Balakrishnan, N., Reddy, R., and Klein-Seetharaman, J. 2008. Transmembrane helix prediction using amino acid property features and latent semantic analysis. BMC Bioinformatics, 9(Suppl. 1), S4, doi: 10.1186/1471-2105-9-S1-S4.

Garcia-Gomez, J. et al. 2005. Corpus based learning of stochastic, context-free grammars combined with Hidden Markov Models for tRNA modelling. International Journal of Bioinformatics Research and Applications, 1(3), 305–18.

Garey, M. and Johnson, D. 1979. Computers and intractability. New York: Freeman, p. 222.

Geschwind, D. and Gregg, J. 2002. Microarrays for the Neurosciences. Cambridge, MA: MIT Press.

Georgiadis, P. et al. 2008. Improving brain tumor characterization on MRI by probabilistic neural networks and non-linear transformation of textural features. Computer Methods and Programs in Biomedicine, 89(1), 24–32.

Ghazavi , S. and Liao, T. 2008. Medical data mining by fuzzy modeling with selected features. Artificial Intelligence in Medicine, doi:10.1016/j.artmed.2008.04.004.

Gianotti , T. et al. 2008. Study of genetic variation in the STAT3 on obesity and insulin resistance in male adults. Obesity, doi: 10.1038/oby.2008.250.

Godsil, C. and Royle, G. 2001. Algebraic graph theory. New York: Springer.

Goffeau, A., et al. 1996. Life with 6000 genes. Science, 274, 546, 563–577.

Gonen, D. et al. 1999. High throughput fluorescent CE-SSCP SNP genotyping. Molecular Psychiatry, 4(4), 339–343.

Guo J., Chen H., Sun Z., and Lin Y. 2004. A novel method for protein secondary structure prediction using dual-layer SVM and profiles. Proteins, 54, 738–743.

Guyon, I., Weston, J., and Barnhill, S. 2002. Gene selection for cancer classification using support vector machines. Machine Learning, 46, 389–422.

Hacia, J. 1999. Resequencing and mutational analysis using oligonucleotide microarray. Nature Genetics, 21, 42–47.

Haldane, J. 1919. The combination of linkage values and the calculation of distance between loci of linked factors. J. Genet., 8, 299–309.

Halushka, M et al. 1999. Patterns of single-nucleotide polymorphisms in candidate genes for blood-pressure homoeostasis. Nature Genetics, 22, 239–247.

Hansen L. and Salamon P. 1990. Neural network ensembles. IEEE Transactions on Pattern Analysis and Machine Intelligence, 12(10), 993–1001.

HapMap. April 2005. www.hapmap.org.

HapMap. October 2006. www.hapmap.org.

Hashibe, M. et al. 2008. Multiple ADH genes are associated with upper aerodigestive cancers. Nature Genetics, 40, 707–709.

Hawley, R. and Walker, M. 2003. Advanced genetic analysis: finding meaning in a genome. Oxford: Blackwell.

Haykin, S. 1994. Neural networks: a comprehensive foundation. Upper Saddle River, NJ: Prentice Hall.

Haykin, S. 1999. Neural networks – a comprehensive foundation, 2nd Ed. Upper Saddle River, NJ: Prentice Hall.

Herbert, A. et al. 2006. A common genetic variant is associated with adult and childhood obesity. Science, 312(5771), 279–283.

Herrero, J., Valencia, A., and Dopzao, J. 2001. A hierarchical unsupervised growing neural network for clustering gene expression patterns. Bioinformatics, 17(2). 126–136.

Hestenes, M. and Stiefel, E. 1952. Methods of conjugate gradients for solving linear systems. Journal of Research of the National Bureau of Standards, 49(6), 409–436.

Heyer, L., Kruglyak, S., and Yooseph, S. 1999. Exploring expression data: identification and analysis of coexpressed genes. Genome Research, 9(11), 1106–1115.

Hill, W. and Robertson, A. 1968. Linkage disequilibrium in finite populations. Theoretical and Applied Genetics, 38, 226–231.

Holland, P. et al. 1991. Detection of specific polymerase chain reaction product by utilizing the 5'in place of 3'exonuclease activity of Thermus aquaticus DNA polymerase. PNAS. USA., 88, 7276–7280.

Hornquist, M., Hertz, J., and Wahde, M. 2003. Effective dimensionality of large-scale expression data using principal component analysis. BioSystem, 65, 147–156.

Hosking, L. K. et al. 2002. Linkage disequilibrium mapping identifies a 390 kb region associated with CYP2D6 poor drug metabolising activity. The Pharmacogenomics Journal, 2, 165–175.

Huang, H., Lee, C., and Ho, S. 2007. Selecting a minimal number of relevant genes from microarray data to design accurate tissue classifiers. Biosystems, 90(1), 78–86.

Huang , S. H., Tan K. K., and Tang, K. Z. 2004. Neural network control: theory and applications. RSP.

Hudson, R. 1987. Estimating the recombination parameter of a finite population model without selection. Genetic Research., 50, 245–250.

Huttenhower , C. et al. 2008. The Sleipnir library for computational functional genomics. Bioinformatics, Advance Access published May 21, 2008.

Hyvärien, A., Karhunen, J., and Oja, E. 2001. Independent component analysis. New York: Wiley.

Ido, P., Oded, M., and Irad, B. 2007. Evaluation of gene-expression clustering via mutual information distance measure. BMC Bioinformatics, 8, 111, doi: 10.1186/1471-2105-8-111.

IHGSC. 2001. Initial sequencing and analysis of the human genome. Nature, 409, 860–921.

International HapMap Consortium. 2005. A haplotype map of the human genome. Nature, 426, 789–796.

International HapMap Consortium. 2004. Integrating ethics and science in the International HapMap Project. 2004. Nature Reviews Genetics., 5, 467–475.

International HapMap Consortium. 2003. The International HapMap Project. Nature, 426, 789–797.

Jeffreys, A., Kauppi, L., and Neumann, R. 2001. Intensely punctuate meiotic recombination in the class II region of the major histocompatibility complex. Nature Genetics, 29, 217–222.

Ji, X. et al. 2003. Mining gene expression data using a novel approach based on Hidden Markov Models. FEBS Letter, 542, 124–131.

Jia, L. and Kitchen, L. 2000. Object-based image similarity computation using inductive learning of contour-segment relations. IEEE Transactions on Image Processing, 9(1), 80–87.

Jiang, D., Pei, J., and Zhang, A. 2003. DHC: a density-based hierarchical clustering method for time series gene expression data. Bioinformatics and Bioengineering, 2003. Proceedings. Third IEEE Symposium on, 10–12 March 2003, pp. 393–400.

Johnson D. 1973. Approximation algorithms for combinatorial problems. Annual ACM Symposium on Theory of Computing, pp. 38–49.

Johnson, G. et al. 2001. Haplotype tagging for the identification of common disease genes. Nature Genetics, 29(2), 233–7.

Johnson, N., Kotz, S., and Balakrishnan, N. 1994. Continuous univariate distributions. Vol. 1, 2nd Ed. New York: Wiley.

Jong, K. 2006. Evolutionary computation: a unified approach. Cambridge MA: MIT Press.

Jutten, C. and Herault, J. 1991 Blind separation of sources, part I: and adaptive algorithm based on neuromimetic architecture. Signal Processing, 24, 1–10.

Kalyanmoy, D. 2004. Optimization for engineering design: algorithms and examples. New Delhi: Prentice-Hall.

Karas, M. and Hillenkamp, E. 1988. Laser desorption ionization of proteins with molecular weight masses exceeding 10,000 Daltons. Analytical Chemistry, 60, 2299–2301.

Karmarkar, N. 1984. A new polynomial-time algorithm for linear programming. Combinatorica, 4, 373–395.

Kauppi, L., Sajantila, A., and Jeffreys, A. 2003. Recombination hotspots rather than population history dominate linkage disequilibrium in the MHC class II region. Human Molecular Genetics, 12, 33–40.

Kearns, M. and Vazirani, U. 1994. Cambridge, MA: MIT Press.

Keedwell, E. and Narayanan, A. 2002. Genetic algorithms for gene expression analysis. First European Workshop on Evolutionary Bioinformatics (2002), 76–86.

Kesseli, J., Ramo, P., and Yli-Harja, O. 2004. Inference of Boolean models of genetic networks using monotonic time transformations. Control, Communications and Signal Processing. First International Symposium on, 21–24 March 2004, pp. 759–762.

Khan, J. et al. 2001. Classification and diagnostic prediction of cancers using gene expression profiling and artificial neural networks. Nature Medicine, 7(6), 673–679.

Kim S., Lee, J., and Bae, J. 2006. Effect of data normalization on fuzzy clustering of DNA microarray data. BMC Bioinformatics, 7, 134, doi: 10.1186/1471-2105-7-134.

Knuth, D. 1997. The art of computer programming, Vol. 1, 3rd Ed. Boston, MA. Addison-Wesley.

Kruglyak, L. and Nickerson, D. 2001. Variation is the spice of life. Nature Genetics, 27, 234–236.

Kwok, P. (eds.). 2002. Single nucleotide polymorphisms: methods and protocols. Totowa, NJ: Humana Press.

Lamers, S. et al. 2008. Prediction of R5, X4, and R5X4 HIV-1 Coreceptor Usage with Evolved Neural Networks. IEEE/ACM Transactions on Computational Biology and Bioinformatics, 5(2), 291–300.

Langers, A. et al. 2008. MMP-2 geno-phenotype is prognostic for colorectal cancer survival, whereas MMP-9 is not. British Journal of Cancer, 98, 1820–1823.

Langmead, C., McClung, C., and Donald, B. 2002. A maximum entropy algorithm for rhythmic analysis of genome-wide expression patterns. Bioinformatics Conference 2002, IEEE. pp. 237–245.

Lawson, C. and Hanson, R. 1974. Solving least-squares problems. Englewood Cliffs, NJ: Prentice-Hall.

Lee, J. and Verleysen, M. 2007. Nonlinear dimensionality reduction. New York: Springer.

Lee, P. and Lee, K. 2000. Genomic analysis. Current Opinion in Biotechnology 11(2), 171–175.

Lee, S. 1984. Multidimensional scaling models with inequality and equality constraints. Communications in Statistics: Simulation and Computation, 13, 127–140.

Levy-Lahad, E. et al. 1995. A familial Alzheimer's disease locus on chromosome 1. Science, 269(5226), 970–973.

Lewontin, R. C. 1964. The interaction of selection and linkage. I. General considerations: heterotic models. Genetics, 49, 49–67.

Liu, B., Cui, Q., Jiang, T., and Ma, S. 2004. A combinational feature selection and ensemble neural network method for classification of gene expression data. BMC Bioinformatics, 5,136, doi:10.1186/1471-2105-5-136.

Liu, C. 1968. Introduction to combinatorial mathematics. McGraw-Hill.

Liu, H. and Hiroshi, M. 1998. Feature selection for knowledge discovery and data mining. Springer.

Liu , K. and Huang, D. 2008. Cancer classification using rotation forest. Computers in Biology and Medicine, 38(5), 601–610.

Liu, Y., Eyal, E., and Bahar, I. 2008. Analysis of correlated mutations in HIV-1 protease using spectral clustering. Bioinformatics, 24(10), 1243–1250.

Lonjou, C. et al. 2003. Linkage disequilibrium in human populations. PNAS, 100, 6069–6074.

Majewski , T. et al. 2008. Understanding the development of human bladder cancer by using a whole-organ genomic mapping strategy. Laboratory Investigation, doi:10.1038/labinvest.2008.27.

Maniatis, N., Collins, A., Xu, C., McCarthy, L., Hewett, D., Tapper, W., Ennis, S., Ke, X., and Morton N. 2002. The first linkage disequilibrium (LD) maps: delineation of hot and cold blocks by diplotype analysis. PNAS, 99(4), 2228–2233.

Maniatis, N., Morton, N., Gibson, J., Xu, C., Hosking, L., and Collins A. 2005. The optimal measure of linkage disequilibrium reduces error in association mapping of affection status. Human Molecular Genetics, 14(1), 145–153.

Maqsood, I., Khan, M., and Abraham, A. 2004. An ensemble of neural networks for weather forecasting. Neural Computation and Applications, 13, 112–122.

Marras, S., Kramer, F., and Tyagi, S. 1999. Multiplex detection of single-nucleotide variations using molecular beacons. Genetics Analysis, 14, 151–156.

Marth G. et al. 1999. A general approach to single-nucleotide polymorphism discovery. Nature Genetics, 23, 453–456.

Martin, S., Zhang, Z., Martino, A., and Faulon, J. 2007. Boolean dynamics of genetic regulatory networks inferred from microarray time series data. Bioinformatics, 23(7), 866–874.

Mattera D. and Haykin S. 1999. Support vector machines for dynamic reconstruction of a chaotic system. Advances in Kernel Methods – Support Vector Learning. Cambridge, MA: MIT Press, 211–242.

MATLAB. 2005. MATLAB documentation: optimization toolbox. MathWorks (online).

McCulloch, W. W. and Pitts, W. 1943. A logical calculus of the ideas imminent in nervous activity. Bulletin of Mathematical Biophysics, 5, 115–133.

Mcshane, L., Radmacher, M., Freidlin, B., Yu, R., Li, M., and Simon, R. 2002. Methods for assessing reproducibility of clustering patterns observed in analyses of microarray data. Bioinformatics, 11, 1462–1469.

Mehrotra, S. 1992. On the implementation of a primal-dual interior point method. SIAM Journal of Optimization, 2, 575–601.

Meng, Z., et al. 2003. Selection of genetic markers for association analysis, using linkage disequilibrium and haplotypes. American Journal of Human Genetics, 73, 115–130.

Miller, C. and Eisenberg , D. 2008. Using inferred residue contacts to distinguish between correct and incorrect protein models. Bioinformatics, Advance Access published May 29, 2008.

Mitchell, T. 1997. Machine learning. New York: McGraw-Hill.

Moore, J., Boczko, E., and Summar, M. 2005. Connecting the dots between genes, biochemistry, and disease susceptibility: systems biology modeling in human genetics. Molecular Genetics and Metabolism, 84(2), 104–111.

Morton, N., Zhang, W., Taillon-Miller, P., Ennis, S., Kwok. P., and Collins, A. 2001. The optimal measure of allelic association. PNAS, 98(9), 5217–5221.

Muller, K., Smola, A., Ratsch, G., Scholkopf, B., Kohlmorgen, J., and Vapnik, V. 1997 Predicting time series with support vector machines. Artificial Neural Networks ICANN'97, Springer, Lecture Notes in Computer Science, 1327, 999–1004.

Murtagh, F. 1983. A survey of recent advances in hierarchical clustering algorithms. Journal of Computation, 26, 354–359.

Murtagh, F. 1984. Complexities of hierarchic clustering algorithms: state of the art. Computational Statistics Quarterly, 1(2), 101–113.

Murtagh, F. 1985. Multidimensional clustering algorithms. COMPSTAT Lectures, 4. Vienna: Physica-Verlag.

Myers, S. et al. 2005. A fine-scale map of recombination rates and hotspots across the human genome. Science, 310(5746), 321–324.

Negoita, M., Neagu, D., and Palade, V. 2005. Computational intelligence: engineering of hybrid systems. Dordrecht: Springer.

Ng, M., Li, M., and Ao, S. et al. 2006. Clustering of SNP data with application to genomics', Proc. 6th IEEE International Conference on Data Mining (ICDM 2006), Hong Kong, 18–22 December 2006, pp. 158–162. IEEE.

NHGRI. 2005. International HapMap Consortium expands mapping effort: map of human genetic variation will speed search for disease genes. NIH (National Institutes of Health) News Release.

Nikkilä, J. et al. 2002. Analysis and visualization of gene expression data using self-organizing maps. Neural Networks, 15(8–9), 953–966.

Nilsson, M. et al. 2001. RNA-templated DNA ligation for transcript analysis. Nucleic Acids Research, 29, 578–581.

Nocedal, J. and Wright, S. 1999. Numeral optimization. New York: Springer.

Ohta, T. and Kimura, M. 1969. Linkage disequilibrium due to random genetic drift. Genetics Research, 13, 47–55.

Oliveira, S. and Seok, S. 2008. A matrix-based multilevel approach to identify functional protein modules. International Journal of Bioinformatics Research and Applications., 4(1), 11–27.

Opitz, D. and Maclin, R. 1999. Popular ensemble methods: an empirical study. Journal of Artificial Intelligence Research, 11, 169–198.

Orita, M et al. 1989. Rapid and sensitive detection of point mutations and DNA polymorphism using the polymerase chain reaction. Genomics., 5, 874–879.

Oto, M. et al. 1993. Optimization of non-radioisotopic single strand conformation polymorphism analysis with a conventional minislab gel electrophoresis apparatus. Analytical Biochemistry, 213(1), 19–22.

Papoulis, A. 1991. Probability, random variables, and stochastic processes, 3rd Ed.. New York: McGraw-Hill.

Pardalos, P. and Resende G. (eds.) 2002. Handbook of applied optimization. Oxford: Oxford University Press.

Patil, N., Berno, A., Hinds, D. Barrett, W., Doshi, J., Hacker, C., Kautzer, C., Lee, D., Marjoribanks, C., McDonough, D. et al. 2001. Blocks of limited haplotype diversity revealed by high-resolution scanning of human chromosome 21. Science, 294, 1719–1723.

Peterson, C. and Ringner, M. 2003. Analyzing tumor gene expression profiles. Artificial Intelligence in Medicine, 28, 59–74.

Pevsner, J. 2003. Bioinformatics and functional genomics. New York: Wiley-Liss.

PLoS. 2005. Genome sequencing: using models to predict who's next. PloS Biology, 3(1), 25.

Principe, J., Euliano, N., and Lefebvre, W. 2000. Neural and adaptive systems: fundamentals through simulations. New York: Wiley.

Prinzie, A. and Van den Poel, D. 2006. Incorporating sequential information into traditional classification models by using an element/position-sensitive SAM. Decision Support Systems, 42(2), 508–526.

Przeworski, M. 2002. The signature of positive selection at randomly chosen loci. Genetics, 162, 2053.

Qin, Z., Niu, T., and Liu, J. 2002. Partitioning-Ligation-Expectation-Maximization Algorithm for haplotype inference with single-nucleotide polymorphisms. American Journal of Human Genetics, 71, 1242–1247.

Quinlan, J. 1990. Learning logical definitions from relations. Machine Learning, 5, 239–266.

Reich, D. E. et al. 2001. Linkage disequilibrium in the human genome. Nature, 411, 199–204.

Reuven, Y., and Zehavit, K. 2004. Approximating the dense set-cover problem. J. Computer and System Sciences. In Press.

Risch, N. and Merikangas, K. 1996. The future of genetic studies of complex human diseases. Science, 273, 1516–1517.

Risch N. J. 2000. Searching for genetic determinants in the new millennium. Nature, 405, 847–856.

Ritchie M. et al. 2001. Multifactor-dimensionality reduction reveals high-order interactions among estrogen-metabolism genes in sporadic breast cancer. Am J Human Genetics, 69(1), 138–47.

Ritchie, M. et al. 2007. Exploring epistasis in candidate genes for rheumatoid arthritis. BMC Proceedings, 1(Suppl. 1), S70.

Rosenblatt, F. 1958. The perceptron: a probabilistic model for information storage and organization in the brain. Psychological Reviews, 65(6), 386–408.

Rowland, J. 2003. Model selection methodology in supervised learning with evolutionary computation. Biosystems, 72(1–2), 187–196.

Rucklidge, W. 1996. Efficient visual recognition using the Hausdorff distance. New York: Springer.

Ryan, D., Nuccie, B., and Arvan, D. 1999. Non-PCR-dependent detection of the factor V Leiden mutation from genomic DNA using a homogeneous invader microtiter plate assay. Molecular Diagnosis, 4, 135–144.

Saeys, Y., Inza, I., and Larranaga, P. 2007. A review of feature selection techniques in bioinformatics. Bioinformatics, 23(19), 2507–2517.

Sajda, P. 2006. Machine learning for detection and diagnosis of disease. Annual Review of Biomedical Engineering, 8, 537–565.

Sakamoto, E. and Iba, H. 2001. Inferring a system of differential equations for a gene regulatory network by using genetic programming. Evolutionary Computation. Proceedings of the 2001 Congress on 27–30 May 2001, 1 720–726.

Salzberg, S. L., Searls, D. B., and Kasif, S. (Eds.). 1998. Computational methods in molecular biology. Amsterdam: Elsevier.

Sanger , F. et al. 1977. Necleotide sequence of bacteriophage phi X194 DNA. Nature, 265, 687–695.

Sawa , T. and Ohno-Machado, L. 2003. A neural network-based similarity index for clustering DNA microarray data. Computers in Biology and Medicine, 33, 1–15.

Scheuner, M., et al. 2004. Contribution of Mendelian disorders to common chronic disease: opportunities for recognition, intervention, and prevention. American Journal of Medical Genetics Part C (Seminars in Medical Genetics), 125C, 50–65.

Sham, P. 1998. Statistics in human genetics. London: Arnold.

Sham, P. and Ao, S. et al. 2007. Combining functional and linkage disequilibrium information in the selection of tag SNPs. Bioinformatics, 23(1), 129–131.

Sherry, S., Ward, M., and Sirotkin, K. 2000. Use of molecular variation in the NCBI dbSNP database. Human Mutation, 15, 68–75.

Shoukri, M., and Pause, C. 1998. Statistical methods for health sciences. Boca Raton, FL: CRC Press.

Singleton, A. et al. 2003. -Synuclein locus triplication causes Parkinson's disease. Science, 302(5646), 841.

Smith, A. et al. 2005. Sequence features in regions of weak and strong linkage disequilibrium. Genome Research, 15, 1519–1534.

Smola, A. and Scholkopf, B. 2004. A tutorial on support vector regression. Statistics and Computing, 14, 199–222.

Spath, H. 1980. Cluster analysis algorithms. Chichester: Ellis Horwood.

Spicker, J. et al. 2002. Neural network predicts sequence of TP53 gene based on DNA chip. Bioinformatics, 18(8), 1133–1134.

Spielman R. S., McGinnis R. E., and Ewen W. J. 1993. Transmission test for linkage disequilibrium: the insulin gene region and insulin-dependent diabetes mellitus (IDDM). American Journal of Human Genetics, 52, 506–516.

Spellman, P. et al. 1998. Comprehensive identification of cell cycle-regulated genes of the yeast saccharomyces cerevisiae by microarray hybridization. Molecular Biology of the Cell, 9, 3273–3297.

Suli, E. and Mayers, D. 2003. An introduction to numerical analysis. Cambridge: Cambridge University Press.

Syeda-Mahmood, T. 2003. Clustering time-varying gene expression profiles using scale-space signals. Bioinformatics Conference, 2003. CSB 2003. Proceedings of the 2003 IEEE, 11–14 Aug. 2003, pp. 48–56.

Tabus, I. and Astola, J. 2003. Clustering the non-uniformly sampled time series of gene expression data. Signal Processing and Its Applications. In Proceedings of the Seventh International Symposium on 1–4 July 2003, 2, 61–64.

Tabus, I., Giurcaneanu, C., and Astola, J. 2004. Genetic networks inferred from time series of gene expression data. Control, Communications and Signal Processing. First International Symposium on, 21–24 March 2004, pp. 755–758.

Taillon-Miller, P. et al. 1999. Efficient approach to unique single nucleotide polymorphism discovery. Genome Research, 9, 499–505.

Taillon-Miller, P. et al. 2000. Juxtaposed regions of extensive and minimal linkage disequilibrium in human Xq25 and Xq28. Nature Genetics, 25, 324–328.

Tan, A. and Gilbert, D. 2003. Ensemble machine learning on gene expression data for cancer classification. Applied Bioinformatics, 2(3 Suppl.), S75–S83.

Taylor, G. et al. 1989. Hypervariable microsatellite for genetic diagnosis. Lancet, 2, 454.

Taylor, G., Day, I. et al. 2005. Guide to mutation detection. New York: Wiley-Liss.

Taylor, J. et al. 2002. Application of metabolomics to plant genotype discrimination using statistics and machine learning. Bioinformatics, 18(Suppl. 2), 241–248.

Theodoridis, S. and Koutroumbas K. 2003. Pattern recognition, 2nd Ed. San Diego, CA: Academic.

Thirion , B. and Faugeras, O. 2003. Dynamical components analysis of fMRI data through kernel PCA. NeuroImage, 20, 34–49.

Thorisson, G. et al. 2005. The International HapMap Project Web site. Genome Research, 15, 1591–1593.

Thornton-Wells, T. et al. 2006. Dissecting trait heterogeneity: a comparison of three clustering methods applied to genotypic data. BMC Bioinformatics, 7, 204.

Tishkoff, S. et al. 2001. Haplotype diversity and linkage disequilibrium at human G6PD: recent origin of alleles that confer malarial resistance. Science, 293(5529), 455–462.

Tome, A. et al. 2007. Greedy kernel PCA applied to single-channel EEG recordings. In Proceedings of the 29th Annual International Conference of the IEEE EMBS, Lyon, France, August 23–26, pp. 5441–5444.

Trumbower, R., Rajasekaran, S., and Faghri, P. 2006. Identifying offline muscle strength profiles sufficient for short-duration FES-LCE exercise: a PAC learning model approach. Journal of Clinical Monitoring and Computing, 20(3), 209–220.

Tyagi, S., Bratu, D., and Kramer, F. 1998. Multicolor molecular beacons for allele discrimination. Nature Biotechnology, 16, 49–53.

Vanteru, B., Shaik, J., and Yeasin, M. 2008. Semantically linking and browsing PubMed abstracts with gene ontology. BMC Genomics, 9(Suppl. 1): S10, doi: 10.1186/1471-2164-9-S1-S10.

Venter, J. et al. 2001. The sequence of the human genome. Science, 291, 1304–1351.

Vogl, T. et al. Accelerating the convergence of the back propagation method. Biological Cybernetics, 59, 257–263.

Vos, P. et al. 1995. AFLP: a new technique for DNA fingerprinting. Nucleic Acids Research, 23, 4407–4414.

Vose, M. 1999. The simple genetic algorithm: foundations and theory. Cambridge, MA: MIT Press.

Wang, D. et al. 1998. Large-scale identification, mapping, and genotyping of single-nucleotide polymorphisms in the human genome. Science, 280, 1077–1082.

Wang, N., Akey, J., Zhang, K., Chakraborty, K., and Jin, L. 2002. Distribution of recombination crossovers and the origin of haplotype blocks: the interplay of population history, recombination, and mutation. American Journal of Human Genetics, 71, 1227–1234.

Wang, S., Yao, J., and Summers, R. 2008. Improved classifier for computer-aided polyp detection in CT colonography by nonlinear dimensionality reduction. Medical Physics, 35(4), 1377–1386.

Wang, X., Li, A., Jiang, Z., and Feng, H. 2006. Missing value estimation for DNA microarray gene expression data by Support Vector Regression imputation and orthogonal coding scheme. BMC Bioinformatics, 7, 32.

Wang, Z., and Moult, J. 2001. SNPs, protein structure and disease. Human Mutation, 17, 263–270.

Waterman M. 1995. Introduction to Computational biology: maps, sequences, and genomes. London/New York: Chapman & Hall.

Weir, B. et al. 2005. Measures of human population structure show heterogeneity among genomic regions. Genome Research, 15, 1468–1476.

Widrow, B. 1959. Generalization and information storage in networks of adaline neurons. Self-organizing systems. Washington, DC: Spartan, pp. 435–461.

Wolkenhauer, O. 2002. Mathematical modeling in the post-genome era: understanding genome expression and regulation-a system theoretic approach. BioSystems, 65, 1–18.

Wu, F., Zhang, W., and Kusalik, A. 2003. Determination of the minimum sample size in microarray experiments to cluster genes using k-means clustering. Bioinformatics and Bioengineering, 2003. In Proceedings of the Third IEEE Symposium on, 10–12 March 2003, pp. 401–406.

Xiao, W. and Oefner, P. 2001. Denaturing high performance liquid chromatography: a review. Human Mutation, 17, 439–474.

Yang, J. 2008. A hybrid machine learning-based method for classifying the Cushing's Syndrome with comorbid adrenocortical lesions. BMC Genomics, 9(Suppl. 1), S23, doi: 10.1186/1471-2164-9-S1-S23.

Yeang, C. and Jaakkola, T. 2003. Time series analysis of gene expression and location data. Bioinformatics and Bioengineering, 2003. In Proceedings of the Third IEEE Symposium on, 10–12 March 2003, pp. 305–312.

Yeung, K. and Ruzzo, W. 2001. Principal component analysis for clustering gene expression data. Bioinformatics, 17(9), 763–774.

Yoshioka, T. and Ishii, S. 2002. Clustering for time-series gene expression data using mixture of constrained PCAS. Neural Information Processing, ICONIP '02, pp. 2239–2243 (v5).

Yukalov, V. 2000. Self-similar extrapolation of asymptotic series and forecasting for time series. Modern Physics Letters B, 14(22/23), 791–900.

Zadeh, L et al. 1996. Fuzzy sets, fuzzy logic, fuzzy systems. Singapore: World Scientific Press.

Zhang, K., Deng, M., Chen, T., Waterman, M., and Sun, F. 2002a. A dynamic programming algorithm for haplotype partitioning. Proceedings of the National Academy of Science, USA, 99, 7335–7339.

Zhang, K. et al. 2002b. Haplotype block structure and its applications to association studies: power and study designs. American Journal of Human Genetics, 71, 1386–1394.

Zhang, K., Qin, Z., Chen, T., Liu, J., Waterman, M., and Sun, F. 2005. HapBlock: haplotype block partitioning and tag SNP selection software using a set of dynamic programming algorithms. Bioinformatics, 21(1), 131–134.

Zhang, L., Zhang, A., and Ramanathan, M. 2003. Fourier harmonic approach for visualizing temporal patterns of gene expression data. Bioinformatics Conference, 2003. CSB 2003. In Proceedings of the 2003 IEEE, 11–14 Aug. 2003, pp. 137–147.